Limnological Conditions and Occurrence of Taste-and-Odor Compounds in Lake William C. Bowen and Municipal Reservoir #1, Spartanburg County, South Carolina, 2006–2009

By Celeste A. Journey, Jane M. Arrington, Karen M. Beaulieu, Jennifer L. Graham, and Paul M. Bradley

Prepared in cooperation with Spartanburg Water, Spartanburg County, South Carolina

Scientific Investigations Report 2011–5060

U.S. Department of the Interior
U.S. Geological Survey

U.S. Department of the Interior
KEN SALAZAR, Secretary

U.S. Geological Survey
Marcia K. McNutt, Director

U.S. Geological Survey, Reston, Virginia: 2011

For more information on the USGS—the Federal source for science about the Earth, its natural and living resources, natural hazards, and the environment, visit http://www.usgs.gov or call 1-888-ASK-USGS

For an overview of USGS information products, including maps, imagery, and publications, visit http://www.usgs.gov/pubprod

To order this and other USGS information products, visit http://store.usgs.gov

Acknowledgments

The authors gratefully acknowledge the assistance of Rebecca West, Ken Tuck, John Moore, and John Westcott with Spartanburg Water in providing technical input and coordination between this investigation and ongoing Spartanburg Water watershed monitoring. Ken Tuck with Spartanburg Water and the staff at the R.B. Simms Water Treatment Plant also provided access and field assistance during sampling activities.

Additionally, Douglas Nagle, U.S. Geological Survey, provided valuable assistance during sampling and data-collection activities. Mark Lowery and Jimmy Clark, U.S. Geological Survey, provided valuable GIS technical support.

Contents

Figures

Tables

Conversion Factors

SI to Inch/Pound

Multiply	By	To obtain
	Length	
centimeter (cm)	0.3937	inch (in.)
millimeter (mm)	0.03937	inch (in.)
meter (m)	3.281	foot (ft)
kilometer (km)	0.6214	mile (mi)
	Area	
hectare (ha)	2.471	acre
hectare (ha)	0.003861	square mile (mi^2)
square kilometer (km^2)	0.3861	square mile (mi^2)
	Volume	
liter (L)	0.2642	gallon (gal)

Inch/Pound to SI

Multiply	By	To obtain
	Length	
foot (ft)	0.3048	meter (m)
mile (mi)	1.609	kilometer (km)
	Area	
acre	4,047	square meter (m^2)
acre	0.4047	hectare (ha)
acre	0.004047	square kilometer (km^2)
square mile (mi^2)	259.0	hectare (ha)
square mile (mi^2)	2.590	square kilometer (km^2)
	Volume	
cubic foot per second (ft^3/s)	0.02832	cubic meter per second (m^3/s)
	Flow rate	
million gallons per day (Mgal/d)	0.04381	cubic meter per second (m^3/s)

Temperature in degrees Celsius (°C) may be converted to degrees Fahrenheit (°F) as follows: °F=(1.8×°C)+32

Abbreviations

ng/L   nanograms per liter

µg/L   micrograms per liter

mL   milliliters

$µm^3/mL$ cubic micrometers per milliliter

mg/d   milligrams per day

mg/L   milligrams per liter

col/mL   colonies per milliliter

Limnological Conditions and Occurrence of Taste-and-Odor Compounds in Lake William C. Bowen and Municipal Reservoir #1, Spartanburg County, South Carolina, 2006–2009

By Celeste A. Journey,[1] Jane M. Arrington,[2] Karen M. Beaulieu,[1] Jennifer L. Graham,[1] and Paul M. Bradley[1]

Abstract

Limnological conditions and the occurrence of taste-and-odor compounds were studied in two reservoirs in Spartanburg County, South Carolina, from May 2006 to June 2009. Lake William C. Bowen and Municipal Reservoir #1 are relatively shallow, meso-eutrophic, warm monomictic, cascading impoundments on the South Pacolet River. Overall, water-quality conditions and phytoplankton community assemblages were similar between the two reservoirs but differed seasonally. Median dissolved geosmin concentrations in the reservoirs ranged from 0.004 to 0.006 microgram per liter. Annual maximum dissolved geosmin concentrations tended to occur between March and May. In this study, peak dissolved geosmin production occurred in April and May 2008, ranging from 0.050 to 0.100 microgram per liter at the deeper reservoir sites. Peak dissolved geosmin production was not concurrent with maximum cyanobacterial biovolumes, which tended to occur in the summer (July to August), but was concurrent with a peak in the fraction of genera with known geosmin-producing strains in the cyanobacteria group. Nonetheless, annual maximum cyanobacterial biovolumes rarely resulted in cyanobacteria dominance of the phytoplankton community.

In both reservoirs, elevated dissolved geosmin concentrations were correlated to environmental factors indicative of unstratified conditions and reduced algal productivity, but not to nutrient concentrations or ratios. With respect to potential geosmin sources, elevated geosmin concentrations were correlated to greater fractions of genera with known geosmin-producing strains in the cyanobacteria group and to biovolumes of a specific geosmin-producing cyanobacteria genus (*Oscillatoria*), but not to actinomycetes concentrations. Conversely, environmental factors that correlated with elevated cyanobacterial biovolumes were indicative of stable water columns (stratified conditions), warm water temperatures, reduced nitrogen concentrations, longer residence times, and high phosphorus concentrations in the hypolimnion. Biovolumes of *Cylindrospermopsis, Planktolyngbya, Synechococcus, Synechocystis*, and *Aphanizomenon* correlated with the greater cyanobacteria biovolumes and were the dominant taxa in the cyanobacteria group.

[1] U.S. Geological Survey.

[2] Spartanburg Water.

Related environmental variables were selected as input into multiple logistic regression models to evaluate the likelihood that geosmin concentrations could exceed the threshold level for human detection. In Lake William C. Bowen, the likelihood that dissolved geosmin concentrations exceeded the human detection threshold was estimated by greater mixing zone depths and differences in the 30-day prior moving window averages of overflow and flowthrough at Lake Bowen dam and by lower total nitrogen concentrations. At the R.B. Simms Water Treatment Plant, the likelihood that total geosmin concentrations in the raw water exceeded the human detection threshold was estimated by greater outflow from Reservoir #1 and lower concentrations of dissolved inorganic nitrogen. Overall, both models indicated greater likelihood that geosmin could exceed the human detection threshold during periods of lower nitrogen concentrations and greater water movement in the reservoirs.

Introduction

Taste-and-odor episodes are common in reservoirs used for drinking water throughout the United States (Paerl and others, 2001; Watson, 2003; Taylor and others, 2006; Zaitlin and Watson, 2006). Cyanobacterial production of trans-1, 10-dimethyl-trans-decalol (geosmin), and 2-methylisoborneol (MIB), which produce musty, earthy tastes and odors in drinking water, represents one of the primary causes of taste-and-odor episodes (Suffet and others, 1996). Additionally, three genera of actinomycetes, a type of bacteria found ubiquitously in soils but that also occurs in the aquatic environment, are important sources of geosmin and MIB (Zaitlin and Watson, 2006). Genera of cyanobacteria, which contain known geosmin and MIB producers, include *Anabaena, Planktothrix, Oscillatoria, Aphanizomenon, Lyngba, Symploca* (Izaguirre and others, 1982; Rashash and others, 1996; Jüttner and Watson, 2007), and *Synechococcus* (Taylor and others, 2006). Genera of actinomycetes that produce geosmin and MIB are *Microbispora, Nocardia*, and *Streptomycetes* (Zaitlin and Watson, 2006; Jüttner and Watson, 2007). Geosmin and MIB are problematic in drinking water because the human taste-and-odor detection threshold for these compounds is extremely low (10 nanograms per liter (ng/L); Wnorowski, 1992; Young and others, 1996; Graham and others, 2008), and conventional water-treatment procedures (particle separation, oxidation, and adsorption) typically do not reduce concentrations below the threshold level (Suffet and others, 1996).

Geosmin and MIB frequently co-occur with cyanotoxins in lakes and reservoirs, though most species of cyanobacteria are not capable of producing taste-and-odor compounds and cyanotoxins simultaneously (Graham and others, 2009, 2010). Taste-and-odor episodes are often sporadic, and intensities vary spatially (Peters and others, 2009). Production and release of cyanotoxins, geosmin, and MIB have been related to cyanobacterial blooms and attributed to environmental factors, including nutrient concentrations and ratios, light availability, water temperatures, water column stability, and flushing rates (Izaguirre and others, 1982; Smith, 1983; Downing and McCauley, 1992; Smith and others, 1995; Smith and Bennett, 1999; Jacoby and others, 2000; Downing and others, 2001; Pearl and others, 2001; Havens and others, 2003; Graham and others, 2004; Dzialowski and others, 2009; Graham and Jones, 2009). Conversely, releases of geosmin and MIB from cyanobacteria also have been associated with periods of high transparency (clear-water phase) attributed to zooplankton grazing (Durrer and others, 1999; Scheffer, 2004; Jüttner and Watson, 2007). However, the complex interaction among the physical, chemical, and biological processes within lakes and reservoirs often makes it difficult to identify primary environmental factors that cause the production and release of these cyanobacterial by-products.

The biological function of these algal-derived compounds is not well known, but production of geosmin and MIB are reported to occur during active growth and extracellular release during stationary periods, cellular senescence, or cell lysis (Rashash and others, 1996). The release of cyanotoxins and taste-and-odor compounds by cyanobacteria simply may be a mechanism for the removal of excess metabolites during periods of environmental stress (Paerl and Millie, 1996; Watson, 2003). The possibility that cyanotoxins, geosmin, and MIB may contribute to the distribution, abundance, and survival of cyanobacteria in the environment has been investigated (Sterner, 1989; Watson, 2003; Scheffer, 2004). Secondary metabolites may deter herbivore grazing and shift grazing pressure toward chemically undefended cyanobacterial and algal species, but that allelopathic role is more often attributed to cyanotoxins (Sterner, 1989). If, however, the availability of chemically undefended algae and cyanobacteria is limited (for example, during seasonally heavy zooplankton grazing events that can produce a clear-water phase), it is possible that a shift could occur in the herbivore community toward species that consume chemically defended cyanobacteria (Scheffer, 2004; Sarnelle and Wilson, 2005; Hansson and others, 2007). Inhibitory effects on the growth of green algae by geosmin have been documented (Ikawa and others, 2001). Conversely, possible bacteriocidal effects of geosmin that may enhance the growth of green algae also have been reported (Sklenar and Horne, 1999; Zaitlin and Watson, 2006).

Background

Lake William C. Bowen (Lake Bowen) and Municipal Reservoir #1 (Reservoir #1) are drinking water supplies in Spartanburg County, South Carolina (fig. 1). Although Lake Bowen and Reservoir #1 are meso-eutrophic and meet State-established nutrient and chlorophyll criteria for lakes and reservoirs, taste-and-odor problems occur periodically (South Carolina Department of Health and Environmental Control, 2006; Journey and Abrahamsen, 2008). Three synoptic surveys were conducted by the U.S. Geological Survey (USGS) in August 2005, May 2006, and October 2006 to evaluate the spatial occurrence of geosmin and MIB and associated limnological conditions in Lake Bowen and Reservoir #1. No periods of high geosmin production were determined during any of these surveys (Journey and Abrahamsen, 2008).

Purpose and Scope

From May 2007 to June 2009, the USGS, in cooperation with Spartanburg Water, conducted a seasonally intensive study of limnological conditions in Lake Bowen and Reservoir #1 to assess the occurrence of cyanobacteria and cyanobacterial-derived compounds, geosmin, MIB, and microcystin. Data collected during the intensive study were merged with the data collected during the 2006 synoptic surveys to enhance the data analysis effort. This report describes the chemical, physical, and biological processes that influence (1) geosmin and microcystin (a common cyanotoxin) occurrence in these source-water reservoirs, (2) cyanobacterial dominance, and (3) geosmin-producing and toxin-producing genera of cyanobacteria. The potential that actinomycetes may be a source of geosmin also was evaluated in this report. An empirical regression model of geosmin concentrations as a function of readily measured, explanatory variables including nutrients, basic water characteristics, and hydrodynamics was developed and can serve as a tool in the ongoing watershed monitoring and management by Spartanburg Water.

Description of the Study Area

Lake Bowen and Reservoir #1 are relatively small, shallow (4.8 and 2.3 meter (m) depth, respectively) cascading impoundments of the South Pacolet River in Spartanburg County, South Carolina (fig. 1; table 1; Journey and Abrahamsen, 2008). At the "full pool" elevation, Lake Bowen has a surface area of 621 hectares (ha) and has 53.2 kilometers (km) of shoreline. Lake Bowen releases spillage (overflow) at the dam directly into Reservoir #1 and by controlled releases at depth from gated conduits (flowthrough). Reservoir #1 is substantially smaller and older than Lake Bowen (table 1). At the full pool elevation, Reservoir #1 has a surface area of 110 ha and 21.1 km of shoreline. Water from these reservoirs is treated by Spartanburg Water at the R.B. Simms Water Treatment Plant. The water-treatment facility and raw water intake are located on Reservoir #1. Recreational activities are allowed on Lake Bowen but are prohibited on Reservoir #1.

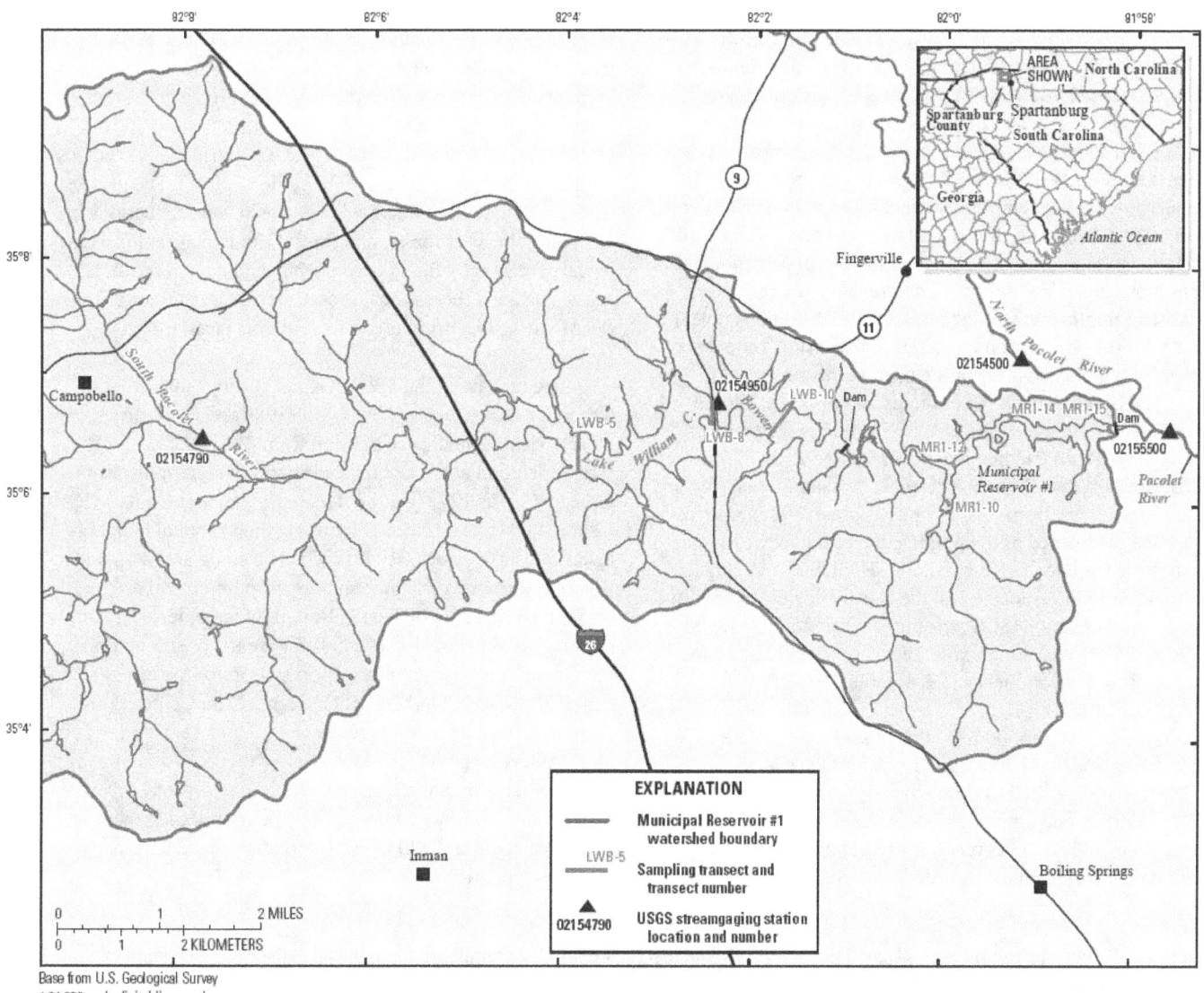

Figure 1. Location of sampling transects in Lake William C. Bowen and Municipal Reservoir #1, Spartanburg County, South Carolina.

Table 1. General physical characteristics of Lake William C. Bowen and Municipal Reservoir #1, Spartanburg County, South Carolina.

[km², square kilometers; ha, hectares; ac, acres; km, kilometer; mi, mile; m, meter; ft, feet]

Characteristic	Units	Lake William C. Bowen	Municipal Reservoir #1
Date of impoundment		1960	1926
Watershed area	km² (mi²)	207 (79.2)	238 (91.5)
Surface area	ha (acres)	621 (1,534)	110 (272)
Shoreline	km (mi)	53 (33)	21 (13.1)
Mean depth	m (ft)	4.8 (15.7)	2.3 (7.5)
Maximum depth	m (ft)	17.4 (57)	13.4 (44)
Mean flushing rate	year	0.60	0.083
Full pool elevation	m (ft)	248.5 (815)	236.6 (777)
Thermal stratification		June to October	June to October
Mean summer mixing zone depth	m (ft)	5.7 (18.7)	5.3 (17.4)
Mean summer euphotic zone depth	m (ft)	4.9 (16.1)	4.8 (15.7)

Sampling sites in Lake Bowen are identified with a prefix of LWB, and sites in Municipal Reservoir #1 are identified with a prefix of MR1 (fig. 1). Site numbers increase in downstream order (for example, LWB-10 is downstream from LWB-8; table 2). This investigation focused on the deeper sites—LWB-10 in Lake Bowen and site MR1-15 in Reservoir #1—but samples were collected at three additional sites (LWB-8 in Lake Bowen and MR1-10 and MR1-12 in Reservoir #1) to evaluate spatial variability of limnological conditions and taste-and-odor occurrence in the reservoirs. Limited samples were collected at sites LWB-5 (3 samples) and MR1-14 (4 samples) in 2007, but these sites were discontinued and the data were not included in this report.

The South Pacolet River drains 143.5 square kilometers (km^2) of the watershed at USGS streamgaging station 02154790 (South Pacolet River near Campobello, SC), which is located less than 1 km upstream from Lake Bowen (fig. 1; table 2). Outflow from Reservoir #1 is about 792 m upstream from the confluence of the South and North Pacolet Rivers that forms the Pacolet River (fig. 1). The USGS streamgaging station 02154500 is located on the North Pacolet River above the confluence, and USGS streamgaging station 02155500 (Pacolet River near Fingerville, SC) is located 322 m downstream from the confluence (fig. 1; table 2). Outflow from Reservoir #1 can be estimated by subtracting the streamflow at gaging station 02155500 from that at station 02154500.

In general, land use within the South Pacolet River Basin in 2001 was classified as rural (Journey and Abrahamsen, 2008; Journey and others, 2011). Forested land dominated the land use in 2001 at 53 percent. Agricultural land use (predominantly hay and pasture) accounted for about 32 percent,

followed by urban land use at more than 11 percent. Urban land use was dominated by low-intensity residential developments, especially around the reservoirs. Recent changes in land use were evaluated by comparing 1992 and 2001 National Land Cover Database (NLCD) geospatial coverages (Homer and others, 2004; Fry and others, 2009). The greatest changes in the South Pacolet River Basin were reductions in agricultural row cropping and forest with concurrent increases in the percentage of land used for hay and pasture, natural grasslands, and recreational grassy areas. Only minor changes (less than 1 percent) in urban land use occurred from 1992 to 2001.

The climate of the Pacolet River Basin is classified as temperate (Kronberg and Purvis, 1959; Purvis and others, 1990). Mean annual precipitation from 1971 to 2000 for the weather station at the Greenville-Spartanburg Airport near Greer, SC, was 127.6 centimeters (cm; National Climatic Data Center, 2004), and the corresponding mean annual temperature was 15.6 degrees Celsius (°C). The study area was under severe to extreme drought conditions from September 2007 through December 2008, as indicated by large departures from the long-term mean (1971–2000) for annual precipitation at the Greenville-Spartanburg Airport station for the calendar years 2007 and 2008 (National Climatic Data Center, 2004). Annual precipitation amounts were 48.7 and 31.0 cm for 2007 and 2008, respectively, and well below the long-term mean precipitation. Annual mean precipitation below long-term mean conditions was common over the past decade, with only 3 years (2003, 2005, and 2009) at or above long-term mean precipitation.

Table 2.   Sampling sites in Lake William C. Bowen and Municipal Reservoir #1, Spartanburg County, and in the Pacolet River basin, Spartanburg County, South Carolina, 2007–2009.

[USGS NWIS, U.S. Geological Survey National Water Information System; ND, no data]

USGS NWIS station number	USGS NWIS station name	Map site identification	Drainage area in square kilometers (square miles)
	Lake Sites		
350624082035200	Lake Bowen below I-26 (Site 5), near Inman, SC	LWB-5	ND
02154950	Lake William C. Bowen at SC Hwy. 9 Bridge near Fingerville, SC	LWB-8	205.1 (79.2)
350641082014700	Lake Bowen below SC Hwy. 9 (Site 10) near Fingerville, SC	LWB-10	ND
3505550820000	Municipal Reservoir 1 at River Oak Road near Fingerville, SC	MR1-10	ND
3506420820154	Municipal Reservoir 1 below Lake Bowen Dam, near Fingerville, SC	MR1-12	ND
02155000	Municipal Reservoir 1 (South Pacolet Reservoir) near Fingerville, SC	MR1-14	238.3 (92)
3506390815814	Municipal Reservoir 1 at Dam near Fingerville, SC	MR1-15	ND
	Tributary Sites		
02154790	South Pacolet River near Campobello, SC	SPAC	143.5 (55.4)
02154500	North Pacolet River near Fingerville, SC	NPAC	300.4 (116)
02155500	Pacolet River near Fingerville, SC	PAC	549.1 (212)

Data Collection and Analysis Methods

Hydrologic, biological, and chemical data were collected during [1]water years (WY) 2007 to 2009 at five sites, and these data were used to evaluate the limnological conditions and phytoplankton community structure in Lake Bowen and Reservoir #1. For this report, an initial assessment of temporal and spatial variations in limnological conditions and phytoplankton community structure was conducted. These assessments were followed by an evaluation of environmental factors associated with geosmin occurrence, cyanobacterial biovolumes, and toxin occurrence. On the basis of the results of the preliminary data analysis, logistic regression models were developed from the data to explain the likelihood of geosmin concentrations exceeding the human detection threshold of 10 ng/L.

Data Collection

Water quality and phytoplankton community structure in Lake Bowen and Reservoir #1 were monitored over a 2-year period (May 2007 to June 2009) to assess the conditions associated with actinomycetes concentrations, cyanobacterial biovolumes, and the occurrences of geosmin, microcystin, and MIB. Fourteen to 16 water samples were collected at two depths, near the surface (1-m depth) and near the bottom (6-m depth, below the thermocline during periods of stratification), along two transects in Lake Bowen (LWB-8, LWB-10) and one transect in Reservoir #1 (MR1-15; fig. 1). Two additional sites in Reservoir #1 (MR1-10 and MR1-12) were sampled near the surface. Samples were collected and processed using USGS protocols and guidelines (U.S. Geological Survey, variously dated; Graham and others, 2008; Graham and others, 2009). Discrete depth samples were collected at three locations across each transect (25, 50, and 75 percent of the width of each transect) using Van Dorn samplers and were mixed in a churn to create a depth-specific composite sample. Transparency (Secchi disk depth) and light attenuation were measured at the time of sampling. Lake profile measurements of fluorescence (an estimate of chlorophyll), specific conductance, pH, dissolved-oxygen concentrations, and water temperature were collected at 1-m depth intervals, including the sampled depths, along each transect using a field-calibrated multiparameter sonde. Reservoir sampling frequency varied seasonally with greater numbers of samples collected during the peak algal growth period (spring to late summer). Lake profile measurements were made more frequently (biweekly compared to monthly) than reservoir samples were collected during the peak algal growth period.

Samples were analyzed for total (dissolved and particulate) nitrogen and total phosphorus as well as for dissolved nitrate plus nitrite, ammonia, orthophosphate, organic carbon, ultraviolet absorbance at 254 and 280 nanometers (estimate of the humic content or reactive fraction of organic carbon), iron, manganese, silica, and major ions by the USGS National Water Quality Laboratory (NWQL) in Denver, Colorado (Fishman and Friedman, 1989; Brenton and Arnett, 1993; Fishman, 1993; American Public Health Association, 1995a, 1995b; 1998; Patton and Kryskalla, 2003). Samples for chlorophyll *a*, pheophytin *a* (pigment degradation product of chlorophyll *a*), and phytoplankton ash-free dry mass were collected on 0.47-micron glass-fiber filters and analyzed according to standard methods and U.S. Environmental Protection Agency (USEPA) method 445.0, respectively (American Public Health Association, 1995b; Arar and Collins, 1997) by the USGS NWQL. For analysis of geosmin and MIB, an aliquot was collected and filtered with a stainless-steel plate-filter assembly and 0.70-micron, 142-millimeter diameter glass-fiber filter into pre-baked amber glass bottles, and analyzed by the USGS Organic Geochemistry Research Laboratory in Lawrence, Kansas, using a gas chromatography and mass spectrometry method with a reporting limit of 0.005 microgram per liter (µg/L; Zimmerman and others, 2002). Raw water samples also were processed and analyzed for dissolved microcystin by the USGS Organic Geochemistry Research Laboratory using an Enzyme-Linked Immunoabsorbent Assay (ELISA) method with a reporting limit of 0.1 µg/L. Samples for the determination of actinomycetes concentration were collected as raw water aliquots in sterile 1-liter plastic bottles and analyzed by the USGS Ohio Microbiology Laboratory, Columbus, Ohio. The double-agar layer (DAL) method, Actinomycete Isolation Agar (AIA), was used for enumeration of actinomycetes (American Public Health Association, 2005).

Additionally, aliquots of raw composited water were used to enumerate and identify phytoplankton. Prior to processing, the samples were agitated in a churn to re-suspend the phytoplankton, and a 250-milliliter (mL) aliquot was removed and preserved in the field with 1 mL of 25-percent glutaraldehyde per 100 mL of sample. Taxonomic characterization and enumeration of phytoplankton in samples were conducted by GreenWater Labs/CyanoLabs of Palatka, Florida. Counts were conducted at multiple magnifications to include organism sizes spanning several orders of magnitude. A minimum of 400 natural units (single cells, colonies, or filaments) per sample were counted for each sample in order to ensure a robust statistical enumeration of the phytoplankton community. Phytoplankton samples were classified at the genus and species level, when possible, with special consideration given to identification of potential geosmin-producing cyanobacteria. Phytoplankton data were analyzed to determine if the algal community structure was dominated by cyanobacteria at the time of sampling. Phytoplankton data were reported as cell density, in cells per milliliter, and as biovolume, in cubic micrometer per milliliter, for each species. Phytoplankton biovolume was calculated by multiplying cell density (the number of cells in a sample [cells per millimeter]) by the volume of each cell (cubic micrometers).

---

[1] A water year extends from October of one calendar year to September of the following calendar year.

Weekly finished and raw water samples were collected by Spartanburg Water personnel at R.B. Simms Water Treatment Plant and analyzed for total geosmin concentrations by a contract laboratory. The period of concurrent weekly total geosmin samples extended from May 2005 to July 2007 and from April 2008 to June 2009. The analytical technique involved a semipermeable membrane extraction method followed by a gas chromatography and mass spectrometry method with a reporting limit of 0.003 μg/L.

Data Analysis

Inflow to Lake Bowen from the South Pacolet River at the USGS streamgaging station 02154790 generally represented 70 to 85 percent of the total inflow into Lake Bowen on the basis of the synoptic stream discharge measurements of minor tributaries in 2008 and 2009 (Journey and others, 2011). Computed unit area discharges (measured synoptic discharge divided by the drainage areas at the measurement site) were similar among the South Pacolet River and its minor tributaries. Therefore, daily mean unit area discharge at the South Pacolet River streamgaging station was multiplied by the total drainage area above Lake Bowen (207 km^2) to extrapolate the total inflow to Lake Bowen for a 5-year period (October 2004 to September 2009). Daily residence times in Lake Bowen were computed by dividing the daily mean inflow to Lake Bowen by the daily mean water volume in Lake Bowen (determined using the stage-volume curve developed by Nagle and others (2009) and the daily mean water level at the USGS streamgaging station 02154950 located at LWB-8). Annual and 30-day moving window average residence times for Lake Bowen were computed from the daily values from October 2004 to September 2009. Outflow from Lake Bowen was considered the major inflow to Reservoir #1 and was computed by the summation of daily volume of water associated with spillage (overflow) and controlled releases. The daily inflow along with annual and 30-day moving window average residence times were calculated for Reservoir #1 as described above for Lake Bowen. Outflow from Reservoir #1 was computed by a 30-day moving window average of differences in streamflow between the Pacolet River USGS streamgaging station (02155500) and the North Pacolet River USGS streamgaging station (02154500).

The degree of stratification was quantified for each sampling event by computing the relative thermal resistance to mixing (RTRM) at 1-m depth intervals from the lake profile of water temperature at the time of sampling at each site. The RTRM for each 1-m depth interval was computed as the ratio of the density difference between water at the top and the bottom of the 1-m depth interval divided by 0.000008 (the density difference between water at 5 and 4 °C; Valentyne, 1957; Wetzel, 2001). The maximum RTRM for that lake profile was used as a measure of the degree of stratification at that site for that sampling event.

Because Lake Bowen potentially could be a major source of geosmin to Reservoir #1, estimates of instantaneous geosmin flux from Lake Bowen to Reservoir #1 were computed from daily overflow and flowthrough volumes and geosmin concentrations (surface and deep) by the equation:

$$F_i = C_i \times V_d \times \text{constant}, \tag{1}$$

where F_i is the instantaneous flux, in milligrams per day; C_i is geosmin concentration at the time of sampling, in nanograms per liter; and V_d is the daily volume of water as overflow and flowthrough, in cubic feet per second. The constant of 2.447 was used to convert the units to milligrams per day. Geosmin concentrations at MR1-12 (the site nearest to Lake Bowen dam) were predicted with a simple mixing model of instantaneous geosmin flux at Lake Bowen divided by the daily volume of water in Reservoir #1 at site MR1-12, in liters.

Water-quality data collected at LWB-8, LWB-10, and MR1-12 from two synoptic surveys conducted in 2006 (Journey and Abrahamsen, 2008) were merged with water-quality data from this investigation and summarized statistically. Water-quality data were censored below the laboratory reporting level (LRL) for several constituents, including geosmin, MIB, nitrate plus nitrite, ammonia, microcystin, and orthophosphate. For data with censored values, descriptive statistics, including mean and median values, were computed by the robust Regression on Order Statistics (ROS) method using the USGS plug-in program incorporated into the Tibco Spotfire S+® 8.1 software (Childress and others, 1999; Helsel, 2005). Median values were used for comparison because they dampen the effects of outliers on the data (Helsel and Hirsch, 1992). Boxplots, based on percentiles of the data distribution, were used as graphical summaries. For example, a 75th percentile represents the concentration below which 75 percent of all measured concentrations occurred. The 50th percentile or median represents the "middle" concentration, such that 50 percent of the data were above and below that concentration. The "box" denotes the interquartile range (from the 75th to the 25th percentile; Helsel and Hirsch, 1992). The "whiskers" display the data range from 90th and 10th percentile, such that 90 percent of the data were between the range outlined by the upper and lower whiskers. Dots represent outliers, data that fall outside of the whisker range.

In general, nonparametric statistical analysis on ranked data was utilized, whereby censored values were given the same rank and ranked below estimated and quantitative (detections above the laboratory reporting limit or LRL) values (Childress and others, 1999; Helsel, 2005). Estimated values that are semiquantitative detections below the LRL were given the same rank, that is, above censored values but below detected values (Childress and others, 1999; Helsel, 2005).

Several exploratory statistical data analyses were applied to the water-quality and phytoplankton data to evaluate the influence of environmental factors on taste-and-odor occurrence. For chemical, physical, and a subset of phytoplankton data, the Kruskal-Wallis test (nonparametric ANOVA on ranked data) was applied to the data to determine if a statistical difference existed (alpha level = 0.05) among groups of data, and the Tukey's Studentized Range test was used to identify which group or groups were different. Results of the Tukey's test were represented by A, B, C where sites with different letters were statistically different from each other (A > B > C), and sites that shared the same letter were statistically indistinct. Initially, water-quality data were evaluated by individual site and depth (1 m and 6 m) for LWB-8, LWB-10, MR1-10, MR1-12, and MR1-15 to determine if water quality differed significantly between reservoirs, among sites, and between depths. Secondly, water-quality data were merged for all sites and depths and evaluated to determine if water quality differed significantly among seasons (winter, January through March; spring, April through June; summer, July through September; fall, October through December). On the basis of findings from the Kruskal-Wallis tests, water-quality data from selected sites (LWB-10 and MR1-15) were evaluated by the Kendall tau correlation procedure to measure the strength of the monotonic bivariate relation between the environmental factors and geosmin concentrations, microcystin concentrations, and cyanobacteria biovolumes (Helsel and Hirsch, 1992). Strong relations between geosmin, cyanobacterial biovolume, and microcystin and a number of potential water-quality, hydrodynamic, and algal drivers were identified and used to help select input variables for regression modeling. Potential water-quality drivers included nutrient links (total nitrogen, dissolved nitrate plus nitrite, dissolved ammonia, total phosphorus, total nitrogen to total phosphorus (TN:TP) ratio, silica, dissolved organic carbon, iron, manganese), basic water characteristics (major ions, water temperature, dissolved oxygen, transparency, pH, euphotic zone depth (Z_{eu}), degree of stratification as measured by RTRM, mixing zone depth (Z_{m})), hydrodynamics (residence times, spillage or overflow, controlled releases), internal phosphorus cycling (difference in total phosphorus concentrations between epilimnion and hypolimnion), and potential sources of geosmin (chlorophyll a, actinomycetes concentration, total phytoplankton biovolume, cyanobacterial biovolume, cyanobacterial dominance, proportion of potential geosmin producers in the cyanobacteria group).

Hypotheses of temporal (seasonal, annual) and spatial (depth, reservoir location) similarities in the taxonomic composition and biovolumes of phytoplankton communities were examined with a series of one-way analysis of similarity (ANOSIM) tests, which are multivariate, nonparametric analogs of analysis of variance (Clarke and Warwick, 2001; Clarke and Gorley, 2006). Higher R-values and lower p-values are evidence of greater dissimilarity; lower R-values and higher p-values indicate relatively similar assemblages (Clarke and Warwick, 2001). Analyses were done using cell biovolumes (in cubic micrometers per milliliter); preliminary analyses of cell densities (cells per milliliter) yielded similar results. Prior to ANOSIM, phytoplankton taxa were aggregated and (or) subset into five taxon by sample data matrices: (1) the major algal divisions, all samples, (2) Cyanophyta genera, all samples (3) known geosmin-producing Cyanophyta genera of Cyanophyta (all seasons, both depths), (4) known geosmin-producing genera of Cyanophyta (spring only, both depths),

and (5) known geosmin-producing genera of Cyanophyta (spring only, surface depths only). Phytoplankton biovolumes were square-root transformed prior to analysis. Four classes of samples for the ANOSIM tests were established: (1) reservoir (Lake Bowen or Municipal Reservoir #1); (2) season (winter—January, February, March; spring—April, May, June; summer—July, August, September; fall—October, November, December); (3) year of collection (water year—October to September; and (4) sample depth (top, 1-m depth, or bottom, 6-m depth).

On the basis of the results of the exploratory statistical analyses, two sites were selected to develop a regression model to estimate geosmin concentrations using environmental factors as explanatory variables. The model could serve as a tool for Spartanburg Water to apply to water-quality and physical data to evaluate the likelihood of geosmin concentrations exceeding the threshold level for human detection. The selected sites were LWB-10 and MR1-15. Site LWB-10 is located near the Lake Bowen dam and represents the quality of water released to Reservoir #1 (fig. 1). Site MR1-15 is located near the Reservoir #1 dam and represents the quality of water near the raw water intake for the R.B. Simms Water Treatment Plant (fig. 1). Because of the high percentage of censored geosmin concentrations at both sites, ordinary least squares regression was not used to develop a multiple linear regression model. Instead, the multiple logistic regression approach was used to identify environmental factors that best explained the likelihood of geosmin concentrations exceeding the human detection threshold of 10 ng/L. In the multiple logistic regression analysis, the response variable was based on a category assigned to geosmin concentrations, in which 1 was assigned to geosmin concentrations above 10 ng/L and 0 was assigned to geosmin concentrations below 10 ng/L. Variables selected as inputs into the multiple logistic regression analysis included those identified in the Kendall tau correlation analysis. Variables were evaluated for normality and transformed accordingly prior to input into the regression model. Logistic regression model equations were developed using the multiple logistic regression routine in SigmaPlot® 11.0. The best equation for each reservoir was selected on the basis of the Pearson Chi-square Statistic (goodness of fit greater at lower statistics and higher p-values), the Hosmer-Lemeshow Statistic (goodness of fit greater at smaller statistics and higher p-values), and the minimum Likelihood Ratio Test Statistic, which tests how well an equation fits the data by summing the squares of the Pearson residuals (goodness of fit greater at lower p-values).

For the model development in Reservoir #1, dissolved geosmin concentrations near the surface (1-m depth) at MR1-15 and total geosmin concentrations in the raw water used by the R.B. Simms Water Treatment Plant were modeled using explanatory factors at the corresponding depth at MR1-15. For the model development in Lake Bowen, dissolved geosmin concentrations near the surface (1-m) at LWB-10 were modeled using explanatory factors at corresponding depth. The model then was applied to data from all depths to evaluate the sensitivity and goodness of fit of the model. Model output provided a Logit P result, whereby Logit P results greater than 0.5 resulted in a positive response (geosmin concentrations exceeded the human detection threshold of 10 ng/L) and less than 0.5 resulted in a reference response (geosmin concentrations were below the human detection threshold of 10 ng/L).

Quality Assurance and Quality Control

Sample collection and processing were conducted according to water-quality sampling and biological assessment protocols documented in the USGS National Field Manual (U.S. Geological Survey, variously dated). A total of 149 water samples were collected from May 2007 to June 2009. Eleven of the 149 samples (about 7 percent) were used for quality assurance for geosmin, MIB, microcystin, nutrient, and major ion analysis. Field blanks were used to test for bias due to contamination during cleaning, collection, processing, and analysis. Blank water was certified free of inorganic (major ions, nutrients, and trace elements) and organic constituents (geosmin, MIB, and microcystin). Field blanks were below the reporting limits for nitrite, dissolved orthophosphate, total phosphorus, and total nitrogen concentrations. Only one field blank had a detectable level of nitrate plus nitrite at an estimated concentration of 0.014 mg/L and ammonia at an estimated concentration of 0.010 mg/L. Two of the 11 field blanks had detectable levels of dissolved orthophosphate at an estimated concentration of 0.003 mg/L and total phosphorus at estimated concentrations of 0.002 and 0.006 mg/L. In general, concentrations in the environmental samples were higher than the detectable levels in the field blanks with the exception of dissolved orthophosphate. Therefore, orthophosphate values were not considered in the data analysis. Geosmin, MIB, or microcystin were not detected in the field blanks.

Limnological Conditions and Taste-and-Odor Occurrence

Computed residence times in Lake Bowen were relatively short and ranged from 0.24 to 1.04 year (table 3). The greatest residence time of 1.04 year occurred in 2008 when drought conditions were prevalent in Spartanburg County. The 5-year (2005–2009) average residence time in Lake Bowen was 0.60 year. A 30-day moving window average of the combined overflow and flowthrough volumes was used in the computation of the Reservoir #1 residence times that ranged from 0.04 (WY 2005) to 0.81 (WY 2008). The 5-year (2005–2009) average residence time in Reservoir #1 was 0.08 year (table 3), almost an order of magnitude less than Lake Bowen. The period of increased residence time was concomitant with the deepest transparency, maximum geosmin concentrations, and maximum cyanobacterial biovolumes in both reservoirs.

Overall, Lake Bowen and Reservoir #1 can be classified as warm monomictic reservoirs that stratify from early June to early-to-late October in locations where depths exceed 5 m

(for detailed bathymetry, see Nagle and others (2009)). During the current study, thermal stratification occurred seasonally in the lower (downstream) regions of the reservoirs at sites LWB-8, LWB-10, and MR1-15. Maximum RTRMs that represented the maximum temperature difference between 1-m depth intervals measured at a site ranged from 2 to 151 at site LWB-8, from 1 to 144 at LWB-10, and from 1 to 185 at MR1-15 (table 4). Maximum RTRMs that were representative of strongly stratified conditions (greater than 80) were prevalent during the summer season, but were near zero from November through the winter season for these sites (fig. 2). Maximum RTRMs were higher during the spring than in the winter, but still represented unstratified conditions (fig. 2; table 4). During periods of stratification, the hypolimnion became anoxic (dissolved-oxygen concentrations below 1 mg/L). Dissolved-oxygen concentrations were statistically lower in the water samples collected below the 6-m depth at LWB-8, LWB-10, and MR1-15 because of the seasonally hypolimnetic anoxia (table 5).

Water column transparencies (Secchi disk depth in meters) were similar among sites in the downstream portion of Lake Bowen (LWB-8, LWB-10) and Reservoir #1 (MR1-15), but decreased in the shallower, upstream portion of Reservoir #1 (MR1-10 and MR1-12; table 5). Median transparencies were between 1.5 and 1.6 m in the downstream regions and 1.0 m in the upstream regions of the reservoirs (table 4). Transparencies also had statistically significant seasonal patterns in Lake Bowen and Reservoir #1, with the greatest transparencies in the spring and the least in the fall (table 6). Site LWB-10 had the greatest seasonal range in transparency ranging from 0.8 to 3.7 m (table 4; fig. 2). Site MR1-15 had a more limited range of 1.1 to 2.0 m (table 4; fig. 2). The pattern of deeper transparencies in the spring is consistent with a "clear water phase" attributed to heavy zooplankton grazing and commonly observed in mesotrophic reservoirs (Durrer and others, 1999; Scheffer, 2004). Zooplankton data, however, were not collected for this study, so the cause of the clear water phase of Lake Bowen or Reservoir #1 could not be verified.

Table 3. Annual and mean annual residence times in Lake William C. Bowen and Municipal Reservoir #1, Spartanburg County, South Carolina, for water years (October–September) 2005–2009.

	Annual mean residence time computed as lake volume/inflow (τ_i) and lake volume/outflow (τ_o)			
	Lake William C. Bowen		**Municipal Reservoir #1**	
Water year	τ_i	τ_o	τ_i	τ_o
2005	0.24	0.22	0.02	0.04
2006	0.45	0.38	0.03	0.13
2007	0.61	0.38	0.04	0.13
2008	1.04	0.68	0.24	0.81
2009	0.68	0.51	0.05	0.16
Average	**0.60**	**0.43**	**0.08**	**0.26**

Table 4. Statistical summary of selected constituents at sites LWB-8 and LWB-10 on Lake William C. Bowen and sites MR1-10, MR1-12, and MR1-15 on Municipal Reservoir #1, South Carolina, August 2005 to June 2009.

[Constituents in bold computed by Regression on Order statistics (Helsel, 1995). ND, no data; E, estimated concentration that is below the laboratory reporting limit; n, number of samples; cen, number of censored values below the laboratory reporting limit; m, meters; mg/L, milligrams per liter; col/mL, colonies per milliliter; µg/L, micrograms per liter; Mgal/d, million gallons per day; Min, minimum; Max, maximum]

Constituents	Units	LWB-8 Surface (May 2006–June 2009)					LWB-8 Bottom (May 2006–June 2009)				
		n (cen)	Mean	Median	Min	Max	n(cen)	Mean	Median	Min	Max
Transparency	m	16 (0)	1.4	1.6	0.8	2.5	NA	NA	NA	NA	NA
Euphotic zone depth	m	13 (0)	5.2	5.4	2.5	7.0	NA	NA	NA	NA	NA
Mixing zone depth	m	14 (0)	6.2	6.5	3.0	9.0	NA	NA	NA	NA	NA
Euphotic:mixing zone ratio	unitless	13 (0)	1.1	1.1	0.5	2.0	NA	NA	NA	NA	NA
Maximum RTRM	unitless	18 (0)	64	54	2	151	NA	NA	NA	NA	NA
Dissolved sulfate	mg/L	16 (0)	2.0	2.0	1.3	2.6	16 (0)	1.9	1.7	1.2	2.7
Dissolved chloride	mg/L	16 (0)	2.9	2.9	2.7	3.2	16 (0)	2.9	2.9	2.7	3.1
Hardness	mg/L as CaCO3	16 (0)	13	13	12	15	16 (0)	13	14	11	16
Dissolved ammonia	**mg/L**	**16 (11)**	**0.039**	**E 0.015**	**<0.02**	**0.220**	**16 (3)**	**0.067**	**0.033**	**<0.02**	**0.237**
Total Kjeldahl nitrogen	mg/L	16 (0)	0.21	0.20	0.09	0.46	16 (0)	0.23	0.22	0.10	0.47
Dissolved nitrate plus nitrite	**mg/L**	**16 (8)**	**0.047**	**0.024**	**<0.016**	**0.158**	**16 (7)**	**0.056**	**0.020**	**<0.016**	**0.177**
Dissolved inorganic nitrogen	**mg/L**	**16 (11)**	**0.089**	**0.049**	**<0.036**	**0.25**	**16 (7)**	**0.127**	**0.101**	**<0.036**	**0.279**
Total organic nitrogen	mg/L	16 (0)	0.19	0.19	0.07	0.26	16 (0)	0.20	0.22	0.04	0.32
Total nitrogen	mg/L	16 (0)	0.29	0.27	0.13	0.49	16 (0)	0.34	0.31	0.23	0.53
Total phosphorus (TP)	mg/L	16 (0)	0.014	0.013	0.010	0.018	16 (0)	0.015	0.015	0.010	0.021
TN:TP ratio	unitless	16 (0)	21	20	11	41	16 (0)	24	22	12	45
Chlorophyll:TP ratio	unitless	16 (0)	0.7	0.6	0.3	1.2	16 (0)	0.7	0.6	0.4	1.4
Actinomycetes	**col/mL**	**11 (4)**	**5.2**	**4**	**<1**	**14**	**10 (2)**	**5.2**	**1.5**	**<1**	**26**
Chlorophyll *a*	µg/L	16 (0)	8.6	7.9	1.8	18.3	16 (0)	10.8	8.7	4.5	26.2
Pheophytin *a*	µg/L	15 (1)	3.6	3.7	<0.10	6.8	15 (0)	6.2	5.6	3.5	12.4
Chlorophyll:pheophytin ratio	unitless	15 (1)	12	2.1	1.7	151	15 (0)	1.8	2.0	0.1	2.6
Dissolved methylisoborneol	**µg/L**	**16 (14)**	**NA**	**<0.005**	**<0.005**	**0.005**	**16 (13)**	**E 0.003**	**E 0.002**	**<0.005**	**0.008**
Dissolved geosmin	**µ**	**16 (7)**	**0.014**	**0.005**	**<0.005**	**0.080**	**16 (7)**	**0.013**	**0.005**	**<0.005**	**0.082**
Total microcystin	**µg/L**	**16 (10)**	**E 0.08**	**E 0.07**	**<0.10**	**0.20**	**14 (13)**	**ND**	**<0.10**	**<0.10**	**0.10**
Total phytoplankton biovolume	µm³/L X 1,000	12 (0)	2,011	2,222	894.0	3,110	12 (0)	2,312	1,947	1,056	4,556
Cyanobacterial biovolume	µm³/L X 1,000	12 (0)	302.0	122.4	62.7	1,577	12 (0)	370.2	120.7	26.66	1,801
Cyanobacteria fraction of total phytoplankton	%	12 (0)	14	8.5	2.5	59	12 (0)	14	5.8	1.8	54
Fraction of known geosmin-producing genera in cyanobacteria group	%	12 (0)	47	50	3.7	74	12 (0)	54	59	7.3	91
30-day prior moving window average residence time (inflow)	days	21 (0)	294	188	100	793	NA	NA	NA	NA	NA
30-day prior moving window average residence time (outflow)	days	21 (0)	172	124	53	388	NA	NA	NA	NA	NA
30-day prior moving window average overflow	Mgal/d	21 (0)	50	60	0	158	NA	NA	NA	NA	NA
30-day prior moving window average flow through	Mgal/d	21 (0)	11.8	2	0	43	NA	NA	NA	NA	NA

Table 4. Statistical summary of selected constituents at sites LWB-8 and LWB-10 on Lake William C. Bowen and sites MR1-10, MR1-12, and MR1-15 on Municipal Reservoir #1, South Carolina, August 2005 to June 2009.—Continued

[Constituents in bold computed by Regression on Order statistics (Helsel, 1995). ND, no data; E, estimated concentration that is below the laboratory reporting limit; n, number of samples; cen, number of censored values below the laboratory reporting limit; m, meters; mg/L, milligrams per liter; col/mL, colonies per milliliter; μg/L, micrograms per liter; Mgal/d, million gallons per day; Min, minimum; Max, maximum]

Constituents	Units	LWB-10 Surface (May 2006–June 2009)					LWB-10 Bottom (May 2006–June 2009)				
		n (cen)	Mean	Median	Min	Max	n(cen)	Mean	Median	Min	Max
Transparency	m	16 (0)	1.8	1.7	0.8	3.7	NA	NA	NA	NA	NA
Euphotic zone depth	m	13 (0)	5.0	5.0	2.5	6.0	NA	NA	NA	NA	NA
Mixing zone depth	m	14 (0)	6.8	7.0	4.0	10.0	NA	NA	NA	NA	NA
Euphotic:mixing zone ratio	unitless	13 (0)	0.8	0.7	0.3	1.3	NA	NA	NA	NA	NA
Maximum RTRM	unitless	18 (0)	73	73	1	144	NA	NA	NA	NA	NA
Dissolved sulfate	mg/L	17 (0)	2 1	2.1	1.3	2.7	12 (0)	2.0	2.1	1.2	2.7
Dissolved chloride	mg/L	17 (0)	2 9	3.0	2.7	3.1	12 (0)	2.9	3.0	2.7	3.1
Hardness	mg/L as CaCO3	20 (0)	13	12	11	15	17 (0)	13	12	12	15
Dissolved ammonia	**mg/L**	**19 (13)**	**0.050**	**0.020**	**<0.020**	**0.225**	**16 (7)**	**0.053**	**0.027**	**<0.020**	**0.232**
Total Kjeldahl nitrogen	mg/L	21 (0)	0 22	0.19	0.09	0.47	17 (0)	0.27	0.25	0.11	0.59
Dissolved nitrate plus nitrite	**mg/L**	**19 (10)**	**0.049**	**0.029**	**<0.016**	**0.147**	**16 (8)**	**0.062**	**0.039**	**<0.016**	**0 16**
Dissolved inorganic nitrogen	**mg/L**	**19 (14)**	**0.099**	**0.081**	**<0.036**	**0.201**	**16 (10)**	**0.106**	**0.069**	**<0.036**	**0 237**
Total organic nitrogen	mg/L	20 (0)	0.19	0.18	0.07	0.28	17 (0)	0.21	0.20	0.05	0 33
Total nitrogen	mg/L	21 (0)	0.31	0.26	0.16	0.53	18 (0)	0.36	0.37	0.23	0.65
Total phosphorus (TP)	mg/L	21 (0)	0.013	0.013	0.008	0.018	18 (0)	0.015	0.013	0.009	0.029
TN:TP ratio	unitless	21 (0)	25	24	14	66	18 (0)	26	22	10	60
Chlorophyll:TP ratio	unitless	20 (0)	0.6	0.6	0.3	1.0	17 (0)	0.8	0.8	0.3	1.4
Actinomycetes	**col/mL**	**12 (3)**	**5**	**4**	**<1**	**18**	**10 (4)**	**5**	**1**	**<1**	**30**
Chlorophyll *a*	μg/L	20 (0)	7.4	7.6	3.2	14.4	17 (0)	12	10.9	3.7	22 9
Pheophytin *a*	μg/L	20 (0)	3.5	3.9	0.8	5.8	17 (0)	6.5	5.1	2.9	13 9
Chlorophyll:pheophytin ratio	unitless	20 (0)	2.4	2.2	1.7	4.0	17 (0)	1.9	2.0	1.3	3.0
Dissolved methylisoborneol	**μg/L**	**20 (16)**	**NA**	**<0.005**	**<0.005**	**0.005**	**17 (12)**	**E 0.004**	**E 0.004**	**<0.005**	**0.006**
Dissolved geosmin	**μ**	**20 (7)**	**0.015**	**0.006**	**<0.005**	**0.100**	**17 (6)**	**0.013**	**0.006**	**<0.05**	**0.070**
Total microcystin	**μg/L**	**20 (13)**	**E 0.08**	**E 0.05**	**<0.10**	**0.30**	**17 (11)**	**E 0.07**	**E 0.06**	**<0.10**	**0 30**
Total phytoplankton biovolume	μm³/L X 1,000	15 (0)	1,926	1,869	3,095	5,951	13 (0)	2,513	2,292	1,222	5,951
Cyanobacterial biovolume	μm³/L X 1,000	15 (0)	265.5	111.8	39.08	1,488	13 (0)	343.9	123.7	43.62	2,205
Cyanobacteria fraction of total phytoplankton	%	15 (0)	13	7.3	2.1	59	13 (0)	12	5	2.4	44
Fraction of known geosmin-producing genera in cyanobacteria group	%	15 (0)	47	51	11	76	13 (0)	55	58	15	90
30-day prior moving window average residence time (inflow)	days	21 (0)	294	188	100	793	NA	NA	NA	NA	NA
30-day prior moving window average residence time (outflow)	days	21 (0)	172	124	53	388	NA	NA	NA	NA	NA
30-day prior moving window average overflow	Mgal/d	21 (0)	50	60	0	158	NA	NA	NA	NA	NA
30-day prior moving window average flow through	Mgal/d	21 (0)	11.8	2	0	43	NA	NA	NA	NA	NA

Table 4.   Statistical summary of selected constituents at sites LWB-8 and LWB-10 on Lake William C. Bowen and sites MR1-10, MR1-12, and MR1-15 on Municipal Reservoir #1, South Carolina, August 2005 to June 2009.—Continued

[Constituents in bold computed by Regression on Order statistics (Helsel, 1995). ND, no data; E, estimated concentration that is below the laboratory reporting limit; n, number of samples; cen, number of censored values below the laboratory reporting limit; m, meters; mg/L, milligrams per liter; col/mL, colonies per milliliter; μg/L, micrograms per liter; Mgal/d, million gallons per day; Min, minimum; Max, maximum]

Constituents	Units	MR1-10 Surface (Sept. 2007 to June 2009)					MR1-12 Surface (May 2007 to June 2009)				
		n(cen)	Mean	Median	Min	Max	n(cen)	Mean	Median	Min	Max
Transparency	m	14 (0)	0.97	1	0.7	1.3	NA	NA	NA	NA	NA
Euphotic zone depth	m	NA	NA	NA	NA	NA	NA	NA	NA	NA	NA
Mixing zone depth	m	NA	NA	NA	NA	NA	NA	NA	NA	NA	NA
Euphotic:mixing zone ratio	unitless	NA	NA	NA	NA	NA	NA	NA	NA	NA	NA
Maximum RTRM	unitless	NA	NA	NA	NA	NA	NA	NA	NA	NA	NA
Dissolved sulfate	mg/L	14 (0)	2.0	2.1	1.3	2.7	14 (0)	2.0	2.2	1.4	2.6
Dissolved chloride	mg/L	14 (0)	2.7	3.0	2.7	3.0	14 (0)	2.9	3.0	2.7	3.1
Hardness	mg/L as CaCO3	14 (0)	13	12	11	14	14 (0)	13	13	11	14
Dissolved ammonia	**mg/L**	**14 (11)**	**0.019**	**E 0.015**	**<0.02**	**0.050**	**16 (13)**	**0.017**	**<0.02**	**<0.02**	**0.113**
Total Kjeldahl nitrogen	mg/L	14 (0)	0.23	0.24	0.12	0.26					
Dissolved nitrate plus nitrite	mg/L	**14 (8)**	**0.034**	**0.019**	**<0.016**	**0.128**	**16 (8)**	**0.037**	**0.018**	**<0.016**	**0.123**
Dissolved inorganic nitrogen	mg/L										
Total organic nitrogen	mg/L										
Total nitrogen	mg/L	14 (0)	0.29	0.29	0.16	0.40	16 (0)	0.29	0.29	0 17	0.40
Total phosphorus (TP)	mg/L	14 (0)	0.020	0.021	0.013	0.028	16 (0)	0.017	0.017	0.011	0.021
TN:TP ratio	unitless	14 (0)	15	13	10	27	16 (0)	19	16	11	44
Chlorophyll:TP ratio	unitless	14 (0)	0.5	0.5	0.3	0.6	16 (0)	0.6	0.6	0.4	1.1
Actinomycetes	**col/mL**	**10 (3)**	**3.5**	**2**	**<1**	**17**	**12 (5)**	**5.1**	**1.6**	**<1**	**40**
Chlorophyll *a*	**μg/L**	14 (0)	9.1	8.7	6.9	14.2	16 (0)	9.1	8.6	7 3	15.5
Pheophytin *a*	μg/L	14 (0)	3.8	3.9	2.6	5	16 (0)	3.7	3.7	2 3	5.4
Chlorophyll:pheophytin ratio	unitless	14 (0)	2.5	2.5	1.7	3.3	16 (0)	2.5	2.4	1.8	3.4
Dissolved methylisoborneol	**μg/L**	**14 (10)**	**E 0.003**	**E 0.002**	**<0.005**	**0.014**	**16 (14)**	**NA**	**<0.005**	**<0.005**	**0.005**
Dissolved geosmin	**μ**	**14 (7)**	**0.006**	**0.005**	**<0.005**	**0.031**	**16 (9)**	**0.009**	**0.004**	**<0.005**	**0.042**
Total microcystin	**μg/L**	**13 (6)**	**E 0.09**	**0.10**	**<0.10**	**0.30**	**16 (11)**	**E 0.07**	**E 0.06**	**<0 10**	**0.20**
Total phytoplankton biovolume	μm³/L X 1,000	11 (0)	2,393	2,439	1,190	3,923	12 (0)	2,742	2,480	1,493	4,619
Cyanobacterial biovolume	μm³/L X 1,000	11 (0)	316.6	175.9	30.43	1,188	12 (0)	344.8	180.0	45 28	1,424
Cyanobacteria fraction of total phytoplankton	%	11 (0)	12	9.3	2.6	33	12 (0)	11	6.9	1.7	31
Fraction of known geosmin-producing genera in cyanobacteria group	%	11 (0)	49	48	9.4	77	12 (0)	49	50	13	74
30-day prior moving window average residence time (inflow)	days	16 (0)	18	13	5	38	16 (0)	18	13	5	38
30-day prior moving window average residence time (outflow)	days	16 (0)	80	43	7	254	16 (0)	80	43	7	254
30-day prior moving window average overflow	Mgal/d	16 (0)	51	58	0	158	16 (0)	51	58	0	158
30-day prior moving window average flow through	Mgal/d	16 (0)	12	1	0	43	16 (0)	12	1	0	43

Table 4. Statistical summary of selected constituents at sites LWB-8 and LWB-10 on Lake William C. Bowen and sites MR1-10, MR1-12, and MR1-15 on Municipal Reservoir #1, South Carolina, August 2005 to June 2009.—Continued

[Constituents in bold computed by Regression on Order statistics (Helsel, 1995). ND, no data; E, estimated concentration that is below the laboratory reporting limit; n, number of samples; cen, number of censored values below the laboratory reporting limit; m, meters; mg/L, milligrams per liter; col/mL, colonies per milliliter; μg/L, micrograms per liter; Mgal/d, million gallons per day; Min, minimum; Max, maximum]

Constituents	Units	MR1-15 Surface (Sept. 2007 to June 2009)					MR1-15 Bottom (Sept. 2007 to June 2009)				
		n(cen)	Mean	Median	Min	Max	n(cen)	Mean	Median	Min	Max
Transparency	m	13 (0)	1.5	1.5	1.1	2.0	NA	NA	NA	NA	NA
Euphotic zone depth	m	13 (0)	5.3	4.9	3.2	7.0	NA	NA	NA	NA	NA
Mixing zone depth	m	14 (0)	7.6	7.5	5.0	11.0	NA	NA	NA	NA	NA
Euphotic:mixing zone ratio	unitless	13 (0)	0.7	0.8	0.4	1.1	NA	NA	NA	NA	NA
Maximum RTRM	unitless	14 (0)	81	77	1	185	NA	NA	NA	NA	NA
Dissolved sulfate	mg/L	16 (0)	2.0	2.1	1.4	2.6	14 (0)	1.9	2.0	1.1	2.7
Dissolved chloride	mg/L	16 (0)	2.9	3.0	2.7	3.1	14 (0)	2.9	3.0	2.7	3.1
Hardness	mg/L as CaCO3	16 (0)	12	12	11	16	14 (0)	13	13	11	16
Dissolved ammonia	**mg/L**	**16 (10)**	**0.031**	**0.019**	**<0.020**	**0.106**	**14 (1)**	**0.061**	**0.052**	**<0.020**	**0.143**
Total Kjeldahl nitrogen	mg/L	16 (0)	0.20	0.22	0.12	0.27	14 (0)	0.20	0.21	0.11	0.24
Dissolved nitrate plus nitrite	**mg/L**	**16 (10)**	**0.032**	**0.017**	**<0.016**	**0.124**	**14 (7)**	**0.03**	**0.014**	**<0.016**	**0.129**
Dissolved inorganic nitrogen	**mg/L**	**16 (11)**	**0.050**	**0.029**	**<0.036**	**0.179**	**14 (7)**	**0.076**	**0.046**	**<0.036**	**0.171**
Total organic nitrogen	mg/L	16 (0)	0.19	0.20	0.07	0.26	14 (0)	0.17	0.17	0.06	0.24
Total nitrogen	mg/L	16 (0)	0.27	0.26	0.18	0.44	14 (0)	0.29	0.28	0.18	0.40
Total phosphorus (TP)	mg/L	16 (0)	0.016	0.015	0.011	0.022	14 (0)	0.016	0.015	0.009	0.025
TN:TP ratio	unitless	16 (0)	18	17	11	31	14 (0)	19	18	11	28
Chlorophyll:TP ratio	unitless	16 (0)	0.6	0.5	0.3	1.1	14 (0)	0.5	0.5	0.3	0.8
Actinomycetes	**col/mL**	**11 (7)**	**3**	**<1**	**<1**	**26**	**11 (2)**	**6**	**3**	**<1**	**31**
Chlorophyll *a*	μg/L	16 (0)	8.9	8.6	3.5	18.5	14 (0)	8.1	7.7	4.7	11.7
Pheophytin *a*	μg/L	16 (0)	3.7	4.0	0.9	6.3	14 (0)	4.1	4.2	2.5	5.2
Chlorophyll:pheophytin ratio	unitless	16 (0)	2.6	2.6	1.5	4.1	14 (0)	2.0	1.9	1.1	3.0
Dissolved methylisoborneol	**μg/L**	**16 (11)**	**NA**	**<0.005**	**<0.005**	**0.005**	**13 (11)**	**NA**	**<0.005**	**<0.005**	**0.005**
Dissolved geosmin	**μ**	**16 (7)**	**0.010**	**0.005**	**<0.005**	**0.050**	**13 (7)**	**0.007**	**E 0.004**	**<0.005**	**0.034**
Total microcystin	**μg/L**	**16 (8)**	**0.10**	**E 0.08**	**<0.10**	**0.40**	**14 (9)**	**E 0.08**	**E 0.07**	**<0.10**	**0.20**
Total phytoplankton biovolume	μm³/L X 1,000	14 (0)	2,025	1,907	900.1	3,684	12 (0)	1,772	1,853	967	2,420
Cyanobacterial biovolume	μm³/L X 1,000	14 (0)	311.8	153.6	30.74	1,615	12 (0)	249.1	127.7	749.1	41.74
Cyanobacteria fraction of total phytoplankton	%	14 (0)	14	6.5	2.5	44	12 (0)	13	6.7	2.8	39
Fraction of known geosmin-producing genera in cyanobacteria group	%	14 (0)	51	55	5.6	83	12 (0)	47	52	10	71
30-day prior moving window average residence time (inflow)	days	16 (0)	18	13	5	38	NA	NA	NA	NA	NA
30-day prior moving window average residence time (outflow)	days	16 (0)	80	43	7	254	NA	NA	NA	NA	NA
30-day prior moving window average overflow	Mgal/d	16 (0)	51	58	0	158	NA	NA	NA	NA	NA
30-day prior moving window average flow through	Mgal/d	16 (0)	12	1	0	43	NA	NA	NA	NA	NA

Table 5. Kruskal-Wallis and Tukey's Studentized Range test results for selected sites and depths in Lake William C. Bowen and Municipal Reservoir #1, South Carolina, May 2007 to June 2009.

[mg/L, milligrams per liter; ft³/s, cubic feet per second; mg/d, milligrams per day; L, liters; μm³/mL, cubic micrometers per milliliter; Alpha level for all tests was 0.05. Results of Tukey's test is represented by A, B, C, where sites with different letters are statistically different from each other (A > B > C) and sites that share the same letter are statistically similar]

Variable	Kruskal-Wallis (Wilcoxon Rank Sum) Test		Tukey's Studentized Range (HSD) Test							
	Chi-squared	p-value	LWB-8-1	LWB-8-6	LWB-10-1	LWB-10-6	MR1-10-1	MR1-12-1	MR1-15-1	MR1-15-6
Transparency (m)	43.50	<0.0001	A	A	A	A	C	B	AB	AB
Dissolved oxygen (mg/L)	27.75	0.000	A	B	A	B	A	A	A	B
Specific conductance (μS/cm)	10.58	0.158	A	A	A	A	A	A	A	A
pH (standard units)	24.16	0.001	AB	B	A	B	AB	AB	AB	B
Dissolved silica (mg/L)	24.31	0.001	AB	A	AB	AB	B	AB	B	AB
Total Kjeldahl nitrogen (mg/L)	9.28	0.233	A	A	A	A	A	A	A	A
Dissolved ammonia (mg/L)	30.09	<0.0001				Differences not identified				
Dissolved nitrate plus nitrite (mg/L)	1.26	0.990	A	A	A	A	A	A	A	A
Total organic nitrogen (mg/L)	8.03	0.330	A	A	A	A	A	A	A	A
Dissolved orthophosphate (mg/L)	2.39	0.935	A	A	A	A	A	A	A	A
Total phosphorus (mg/L)	27.12	0.000	B	AB	B	B	A	AB	AB	AB
Total nitrogen (mg/L)	12.10	0.097	A	A	A	A	A	A	A	A
Total nitrogen:total phosphorus ratio (unitless)	19.42	0.007	AB	AB	A	A	B	AB	AB	AB
Chlorophyll a:total phosphorus ratio (unitless)	19.05	0.008	AB	AB	AB	A	B	AB	AB	B
Actinomycetes (col/mL)	5.14	0.643	A	A	A	A	A	A	A	A
Chlorophyll a (μg/L)	9.76	0.202	A	A	A	A	A	A	A	A
Pheophytin a (μg/L)	24.50	0.001	B	A	B	A	B	B	B	B
Chlorophyll a:pheophytin a ratio (unitless)	25.07	0.001				Differences not identified				
Dissolved methylisoborneol (μg/L)	3.81	0.802	A	A	A	A	A	A	A	A
Total microcystin (μg/L)	3.34	0.852	A	A	A	A	A	A	A	A
Dissolved geosmin (μg/L)	4.24	0.752	A	A	A	A	A	A	A	A
Total phytoplankton biovolume (μm³/mL)	8.73	0.273	A	A	A	A	A	A	A	A
Cyanobacteria (blue-green algae) biovolume (μm³/mL)	2.26	0.944	A	A	A	A	A	A	A	A
Fraction of cyanobacteria in total phytoplankton (by biovolume; unitless)	1.15	0.992	A	A	A	A	A	A	A	A
Fraction of geosmin producers in cyanobacteria biovolume (unitless)	2.60	0.919	A	A	A	A	A	A	A	A

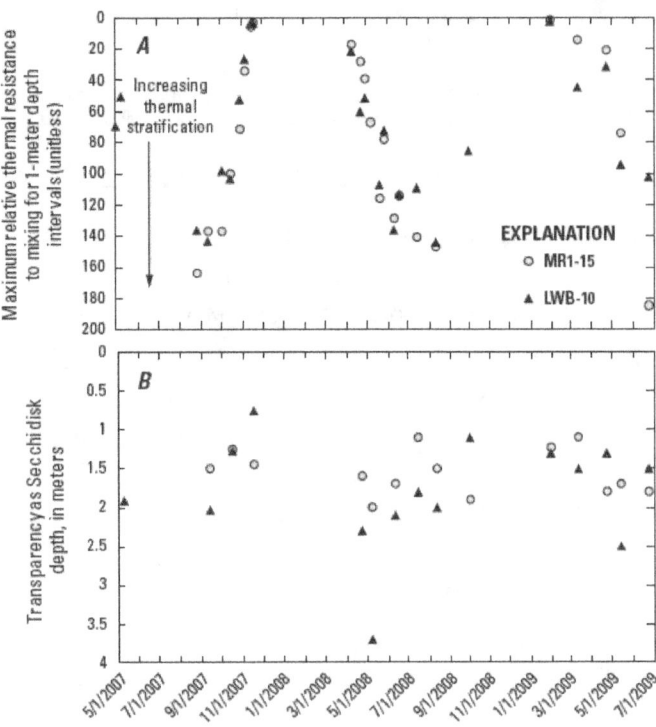

Figure 2. Temporal variation in maximum relative thermal resistance to mixing (RTRM) computed for 1-meter depth intervals as an indicator of degree of stratification and transparency as Secchi disk depth at site LWB-10 in Lake William C. Bowen and site MR1-15 in Municipal Reservoir #1, Spartanburg County, South Carolina, May 2007 to June 2009 [Maximum RTRM values greater than 80 were considered to represent strongly stratified conditions.]

General Limnological Conditions

Maximum nutrient concentrations remained below 0.034 mg/L for total phosphorus and 0.70 mg/L for total nitrogen in Lake Bowen and Reservoir #1 for the study period. These values are well below the South Carolina Department of Health and Environmental Control (SCDHEC) numeric nutrient criteria for lakes and reservoirs in the Piedmont ecoregion of South Carolina (0.060 mg/L for total phosphorus and 1.50 mg/L for total nitrogen; South Carolina Department of Health and Environmental Control, 2006; table 4; fig. 3). During the study period, the maximum chlorophyll *a* concentration was 26 µg/L at LWB-8, but median concentrations were below 9 µg/L at all sites in both reservoirs (table 4). All chlorophyll *a* concentrations were below the SCDHEC numeric chlorophyll *a* criterion of 40 µg/L (table 4; fig. 4). Median chlorophyll-to-total-phosphorus ratios ranged from 0.5 to 0.7 for all sites in Lake Bowen and Reservoir #1, indicating phosphorus limiting conditions (table 4; fig. 5). Overall,

Lake Bowen and Reservoir #1 had statistically similar water chemistry, with nutrient concentrations indicative of mesotrophic conditions (Nürnberg, 1996; table 5). Median chlorophyll *a* concentrations in both reservoirs were indicative of mesotrophic conditions; however, maximums were indicative of eutrophic conditions. On the basis of TN:TP ratios (10:66), potential phosphorus limitation (TN:TP >17) or co-limitation (17 < TN:TP > 10) by phosphorus and nitrogen (Forsberg and Ryding, 1980) was common in these reservoirs. Low total-nitrogen-to-total-phosphorus (TN:TP) ratios (generally below 29:1) were consistent with environmental conditions reported to favor nitrogen-fixing cyanobacteria (Smith, 1983; Havens and others, 2003; fig. 4).

Nutrient and trophic indicators were evaluated among sites and depths and between reservoirs to identify any significant differences. Nitrogen species had statistically similar concentrations among sites and between reservoirs (table 5). Total phosphorus concentrations were statistically similar at all sites except MR1-10 (shallow, upper region of Reservoir #1), which had statistically greater total phosphorus concentrations than most sites and depths in Lake Bowen (fig. 3; table 5). Chlorophyll *a* concentrations also were similar among sites and depths in both reservoirs. Conversely, pheophytin *a* concentrations were higher below the 6-m depth at sites LWB-8 and LWB-10 than at the 1-m depth for those sites, which produced lower chlorophyll *a*:pheophytin *a* ratios at the 6-m depths at sites LWB-8 and LWB-10 (Tukey's test was unable to identify statistical differences; table 5). In general, nutrient and trophic indicators were similar between reservoirs and among sites and depths, especially when comparing sites in the downstream regions of both reservoirs (LWB-8, LWB-10, and MR1-15). Therefore, site LWB-10 in Lake Bowen was selected to represent the quality of water entering Reservoir #1, and MR1-15 in Reservoir #1 was selected to represent the quality of water entering the treatment plant drinking-water intake. The two selected sites were incorporated into further statistical analysis to determine which environmental factors influence geosmin concentrations and cyanobacterial biovolumes in these reservoirs.

Seasonal variation in dissolved inorganic nitrogen (dissolved nitrate plus nitrite and ammonia) concentrations indicated periods of limitation and enrichment in readily bioavailable forms of nitrogen for most phytoplankton groups. Dissolved inorganic nitrogen (DIN) concentrations were consistently low in the reservoirs near the surface (1 m) during the spring and summer of 2008, but were elevated relative to the surface concentrations at the bottom depth (6 m) at MR1-15 (fig. 5). The elevated DIN concentrations near the bottom were attributed to increased dissolved ammonia in the anoxic hypolimnion. Kruskal-Wallis tests were applied to the combined 2006-to-2009 water-quality data of all depths and sites to identify seasonal variations in these data (table 6). Dissolved ammonia concentrations were highest in the fall and lowest in the spring and summer, and dissolved nitrate plus nitrite concentrations were highest in the winter and spring and lowest in the summer and fall (table 6). Increases in total phosphorus

Table 6. Kruskal-Wallis and Tukey's Studentized Range test results for selected environmental variables by season for sites in Lake William C. Bowen and Municipal Reservoir #1, Spartanburg County, South Carolina, May 2007 to June 2009.

[HSD, honestly significant difference; mg/L, milligrams per liter; ft³/s, cubic feet per second; mg/d, milligrams per day; L, liters; μm³/mL, cubic micrometers per milliliter; Results of Tukey's test are represented by A, B, C, where seasons with different letters are statistically different from each other (A > B > C) and seasons that share the same letter are statistically similar]

Variable	Kruskal-Wallis (Wilcoxon Rank Sum) Test		Tukey's Studentized Range (HSD) Test			
	Chi-squared	p-value	Summer	Spring	Fall	Winter
Transparency (m)	25.70	<0.0001	AB	A	C	BC
Euphotic zone depth (m)	23.17	<0.0001	A	A	B	AB
Mixing zone depth (m)	46.25	<0.0001	C	B	B	A
Euphotic:mixing zone depth ratio (unitless)	38.18	<0.0001	A	B	BC	C
Maximum relative thermal resistance mixing (1-m intervals)	51.06	<0.0001	A	B	BC	C
Dissolved silica (mg/L)	47.74	<0.0001	B	C	A	A
Total Kjeldahl nitrogen (mg/L)						
Dissolved ammonia (mg/L)	25.21	<0.0001	B	B	A	AB
Dissolved nitrate plus nitrite (mg/L)	86.18	<0.0001	C	B	C	A
Total organic nitrogen (mg/L)	13.87	0.003	B	A	B	AB
Dissolved orthophosphate (mg/L)	32.74	<0.0001	A	B	B	B
Total phosphorus (mg/L)	9.76	0.021	Not determined			
Chlorophyll *a* (μg/L)	7.63	0.054	A	A	A	A
Pheophytin *a* (μg/L)	0.68	0.878	A	A	A	A
Methylisoborneol (μg/L)	66.75	<0.0001	B	B	B	A
Dissolved microcystin (μg/L)	8.84	0.031	Not determined			
Dissolved geosmin (μg/L)	38.08	<0.0001	B	A	B	AB
Change in total phosphorus (Hypolimnion TP - Epilimnion TP) (mg/L)	20.51	0.000	A	B	B	B
Total phytoplankton biovolume (μm³/mL)	8.50	0.037	A	AB	B	AB
Cyanobacteria (blue-green algae) biovolume (μm³/mL)	53.73	<0.0001	A	B	B	B
Fraction of cyanobacteria in total phytoplankton (by biovolume; unitless)	56.61	<0.0001	A	B	B	B
Total nitrogen:total phosphorus ratio (unitless)	29.06	<0.0001	B	A	A	A
Total nitrogen (mg/L)	24.81	<0.0001	B	A	AB	A

concentrations in the hypolimnion relative to the epilimnion during summer to early fall were evident at sites LWB-10 and MR1-15, especially during the summer of 2008 (fig. 5). Hypolimnetic increase in total phosphorus concentrations was concurrent with the development of strong stratification (generally at maximum RTRMs greater than 100) and anoxic conditions in the summer and early fall, indicating a source of phosphorus from the bed sediment during that seasonal period (figs. 2, 5). Hypolimnetic increases in total phosphorus concentrations were as high as 0.017 mg/L at LWB-10 and 0.012 mg/L at MR1-15. Except for dissolved nitrate plus nitrite, total nitrogen and TN:TP ratios were statistically lower during the summer. For all depths and sites, total organic nitrogen concentrations were statistically greater during the spring and winter than during the summer and fall.

Phytoplankton Community Structure

Median total phytoplankton biovolumes ranged from 1,853,000 (site MR1-15 bottom depth) to 2,480,000 cubic micrometers per milliliter (μm³/mL) (MR1-12) in Reservoir #1 and from 1,869,000 (site LWB-10 surface depth) to 2,292,000 μm³/mL (site LWB-10 bottom depth) in Lake Bowen (table 4). Cyanobacterial biovolumes were an order of magnitude lower, with median biovolumes ranging from 127,700 (MR1-15 bottom depth) to 180,000 μm³/mL (MR1-12) in Reservoir #1 and from 111,800 (LWB-10 surface depth) to 123,700 μm³/mL (site LWB-10 bottom depth) in Lake Bowen (table 4; fig. 6). Overall, no statistical differences were identified in total phytoplankton and cyanobacterial biovolumes among sites and depths in Lake Bowen and Reservoir #1 (table 5). Maximum cyanobacterial biovolumes

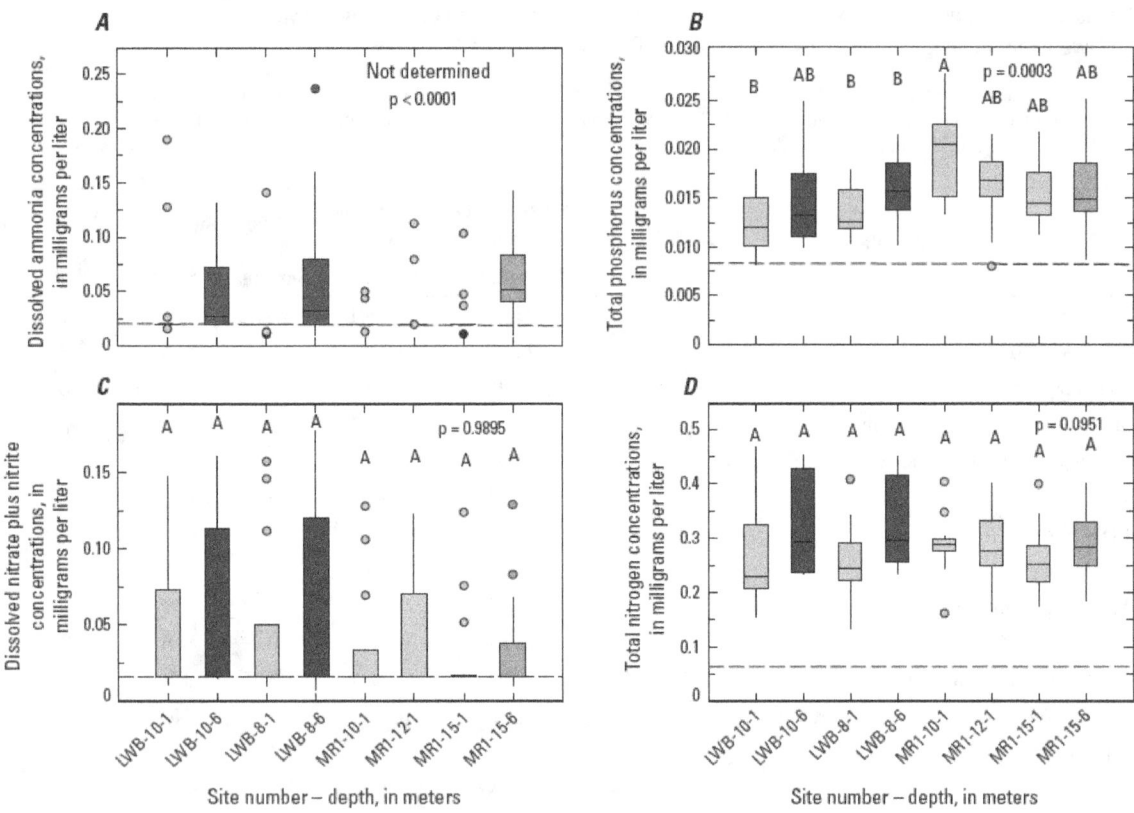

Figure 3.   Range in *(A)* dissolved ammonia, *(B)* total phosphorus concentrations, *(C)* nitrate plus nitrite, and *(D)* total nitrogen concentrations in Lake William C. Bowen and Municipal Reservoir #1, Spartanburg County, South Carolina. Results of the Tukey test is represented by the probability value (p) and by A, B, C, where sites with different letters are statistically different from each other (A > B > C) and sites that share the same letter are statistically similar. Horizontal lines on dissolved ammonia and nitrate plus nitrite plots represent the laboratory reporting level.

occurred during the summer (July to August), but did not always result in cyanobacteria dominating the phytoplankton community (table 6). In fact, median percentages of the cyanobacterial fraction of the total phytoplankton community were less than 10 percent for all sites, indicating cyanobacteria rarely dominated the phytoplankton community (table 4). The exception occurred in July 2008 in Lake Bowen, when cyanobacteria represented as much as 59 percent of the total phytoplankton community in Lake Bowen (fig. 7C). During the same time period, cyanobacteria dominance was less pronounced in Reservoir #1 (31 to 44 percent of the total phytoplankton community) than Lake Bowen (fig. 7C). Additionally, surface algal blooms were not observed during the study period.

Within the cyanobacteria group, genera that contained known geosmin-producing species and known toxin-producing species were present at all sites. Biovolumes of these genera, however, varied seasonally and annually. Known geosmin-producing genera identified in the two reservoirs included *Planktolyngbya, Aphanizomenon Synechococcus, Psuedoanabaena, Oscillatoria,* and *Anabaena,* many of which can also produce toxins. Known toxin-producing genera

identified in the two reservoirs included *Cylindrospermopsis, Synechocystis, Microcystis,* and *Aphanacapsa.* Median percentages of known geosmin-producing genera in the cyanobacterial group ranged from 48 to 59 percent among all sites (table 4). Fractions of all known geosmin-producing genera in the cyanobacterial group were similar among sites and depths (table 5). However, seasonal differences were identified that indicated greater fractions of known geosmin-producing genera occurred during the spring and winter and the least fractions occurred during the fall (fig. 7; table 6).

Results of the ANOSIM tests for the differences among samples grouped by reservoir, season, year, and depth position indicated significant and relatively strong temporal variation (particularly seasonal variation) in phytoplankton assemblages in terms of biovolumes of algal divisions, all Cyanophyta genera, and known geosmin-producing Cyanophyta genera (table 7). The greatest seasonal differences were for known geosmin-producing Cyanophyta genera (including all seasons and all depths) during the spring. Conversely, taxonomic assemblages were not different between Lake Bowen and Reservoir #1 or between depths (table 7). Once the taxonomic assemblage of known geosmin-producing Cyanophyta genera

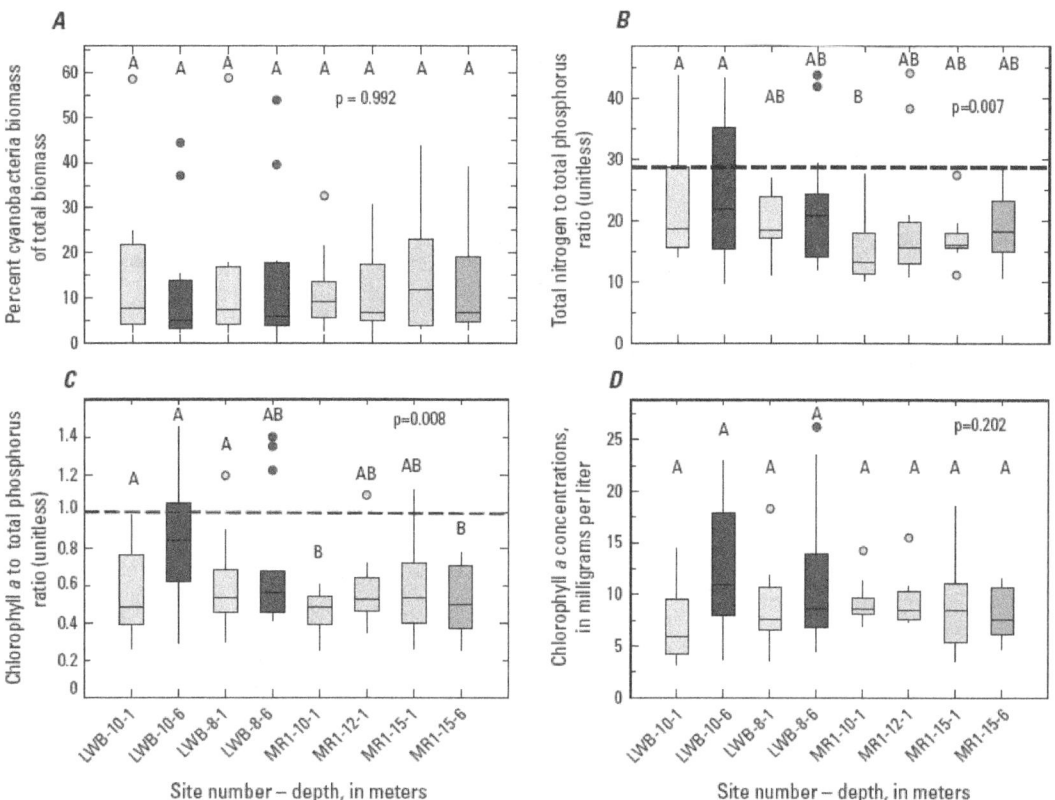

Figure 4. Range in percentage of cyanobacteria in *(A)* total phytoplankton biomass, *(B)* total nitrogen to total phosphorus ratios, *(C)* chlorophyll *a* to total phosphorus ratios, and *(D)* chlorophyll *a* concentrations for two depths in Lake William C. Bowen and Municipal Reservoir #1, Spartanburg County, South Carolina. [Results of the Tukey test is represented by A, B, C, where sites with different letters are statistically different from each other (A > B > C) and sites that share the same letter are statistically similar. Total nitrogen to total phosphorus ratios below the dashed line represent conditions favorable for nitrogen-fixing cyanobacteria. Chlorophyll *a* to total phosphorus ratios well below the dashed line are indicative of phosphorus-limiting conditions.

became restricted by season (spring only) and depth position (spring only, surface depth only), no differences among samples were identified (table 7). In summary, multivariate statistical analysis indicated that taxonomic assemblages of the phytoplankton community (represented by the major algal divisions) and the cyanobacterial community varied season to season and year to year. Conversely, the above-mentioned taxonomic assemblages between the two reservoirs and between the two depth positions were similar. This pattern of variability is consistent with the pattern identified in the Kruskal-Wallis analysis of the chemical and physical data (tables 5, 6).

Geosmin, MIB, and Microcystin Occurrence

Geosmin was the most commonly detected taste-and-odor compound in Lake Bowen and Reservoir #1 during the study period from May 2006 to June 2009 (Journey and Abrahamsen, 2008). However, about 35 percent of the samples at

LWB-10 and more than 44 percent of the samples at MR1-15 had geosmin concentrations below the LRL (0.005 μg/L). Median geosmin concentrations ranged from 0.004 to 0.006 μg/L at the study sites (fig. 6; table 4). MIB rarely was detected (median concentrations were less than the laboratory reporting level of 0.005 μg/L at all sites; fig. 6; table 4). When present, MIB was at very low concentrations with maximum MIB concentrations ranging from 0.005 (LWB-10 surface depth, LWB-8 surface depth, MR1-12, MR1-15 surface and bottom depths) to 0.014 μg/L (MR1-10; table 4). As is often observed in surface-water systems, geosmin concentrations in Lake Bowen and Reservoir #1 exhibited strong annual and seasonal variability during the study (fig. 7). Maximum geosmin concentrations occurred in April and May 2008 at all sites in both reservoirs (fig. 7; table 4). During the peak geosmin period in the spring of 2008, maximum geosmin concentrations ranged from 0.082 to 0.100 μg/L in Lake Bowen, which were 8 to 10 times greater than the human detection threshold

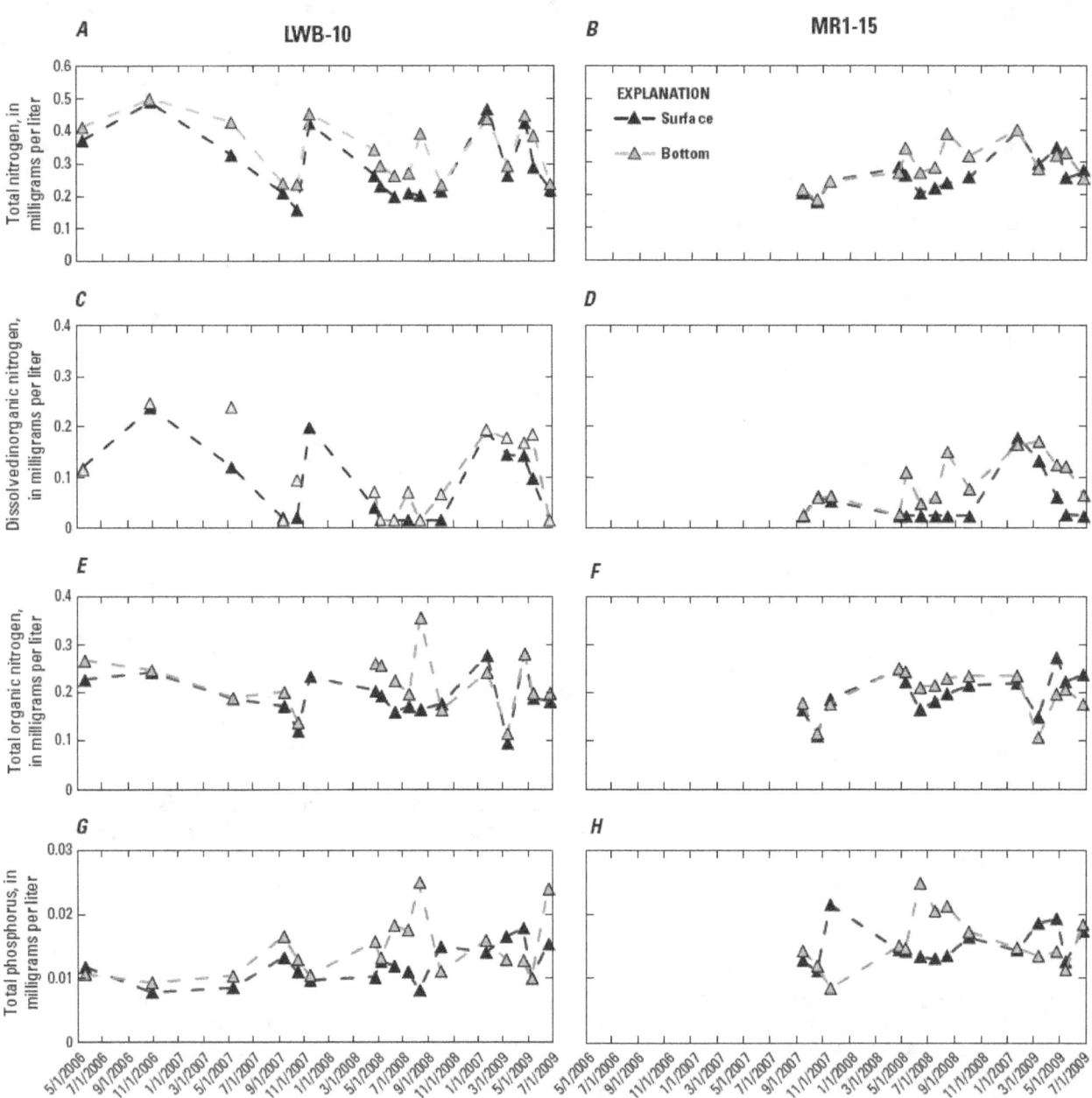

Figure 5.   Temporal variation in the epilimnetic (surface) and hypolimnetic (bottom) concentrations of *(A)* total nitrogen *(C)* dissolved inorganic nitrogen *(E)* total organic nitrogen and *(G)* total phosphorus at site LWB-10 in Lake William C. Bowen and of *(B)* total nitrogen, *(D)* dissolved inorganic nitrogen, *(F)* total organic nitrogen and *(H)* total phosphorus site MR1-15 in Municipal Reservoir #1 Spartanburg County South Carolina, May 2006 to June 2009.

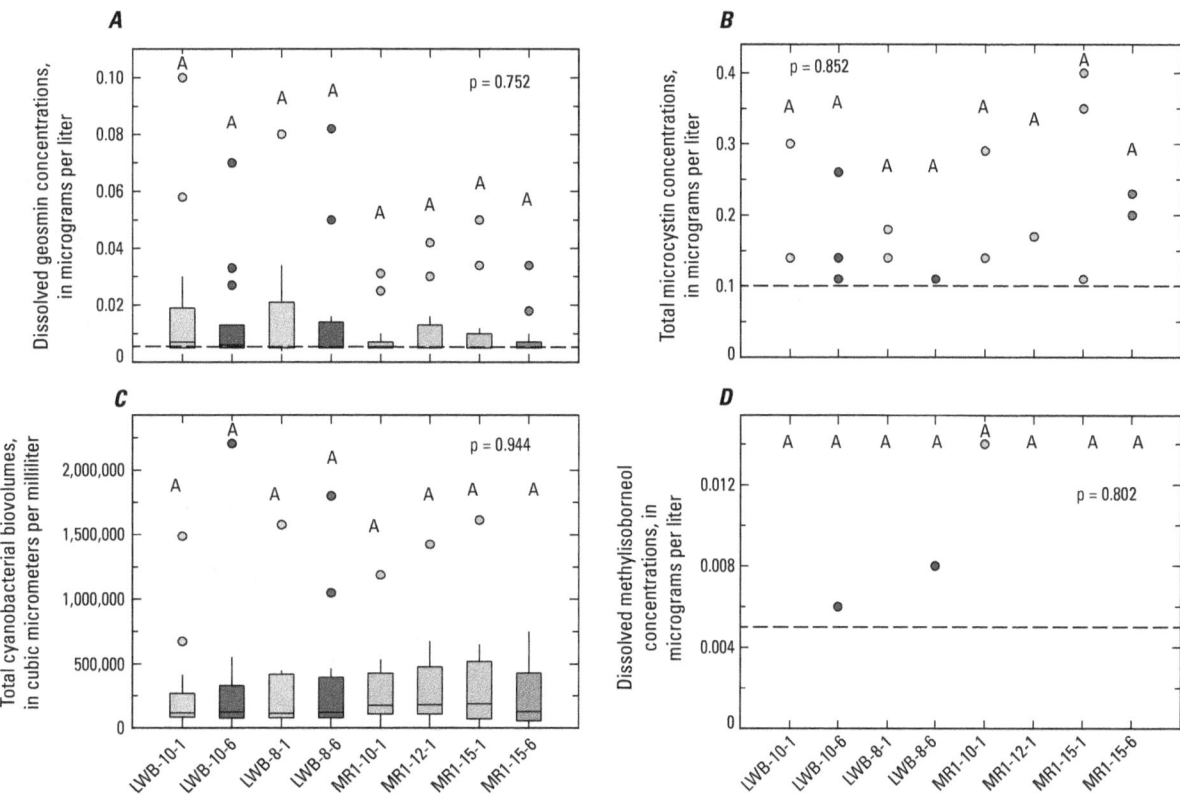

Figure 6. Range in *(A)* dissolved geosmin, *(B)* total microcystin, and *(D)* dissolved methylisoborneol concentrations and *(C)* cyanobacterial biovolumes in Lake William C. Bowen and Municipal Reservoir #1, Spartanburg County, South Carolina. Results of the Tukey test are represented by the probability value (p) and by A, B, C, where sites with different letters are statistically different from each other (A > B > C) and sites that share the same letter are statistically similar. Horizontal dashed lines on plots represent the laboratory reporting level.

Table 7. Nonparametric Analysis of Similarity (one-way) results for five phytoplankton taxonomic assemblages in relation to four factors of reservoir group, season, year, and depth position for Lake William C. Bowen and Municipal Reservoir #1, Spartanburg County, South Carolina, May 2007 to June 2009.

Phytoplankton taxonomic assemblages	Analysis of Similarity (Nonparametric One-Way ANOSIM)							
	Reservoir Group		**Season**		**Year**		**Depth Position**	
	Global R	p-value	Global R	p-value	Global R	p-value	Global R	p-value
Algal Division								
Biovolume	0.009	0.740	0.401	0.001	0.231	0.001	0.078	0.006
Cell Density	0.078	0.055	0.349	0.001	0.27	0.001	0.045	0.006
Cyanophyta genera								
Biovolume	0.005	0.027	0.723	0.001	0.331	0.001	0.045	0.051
Cell Density	0.029	0.036	0.633	0.001	0.358	0.001	0.024	0.18
Known geosmin-producing Cyanophyta genera (all seasons, all depths)								
Biovolume	0.017	0.110	0.595	0.001	0.280	0.001	0.086	0.01
Cell Density	0.026	0.029	0.591	0.001	0.333	0.001	0.056	0.073
Known geosmin-producing Cyanophyta genera (spring only, all depths)								
Biovolume	0.051	0.056	NA	NA	0.146	0.017	0.22	0.003
Cell Density	0.099	0.012	NA	NA	0.154	0.011	0.187	0.001
Known geosmin-producing Cyanophyta genera (spring only, surface depths only)								
Biovolume	0.03	0.236	NA	NA	0.113	0.083	NA	NA
Cell Density	0.047	0.150	NA	NA	0.063	0.200	NA	NA

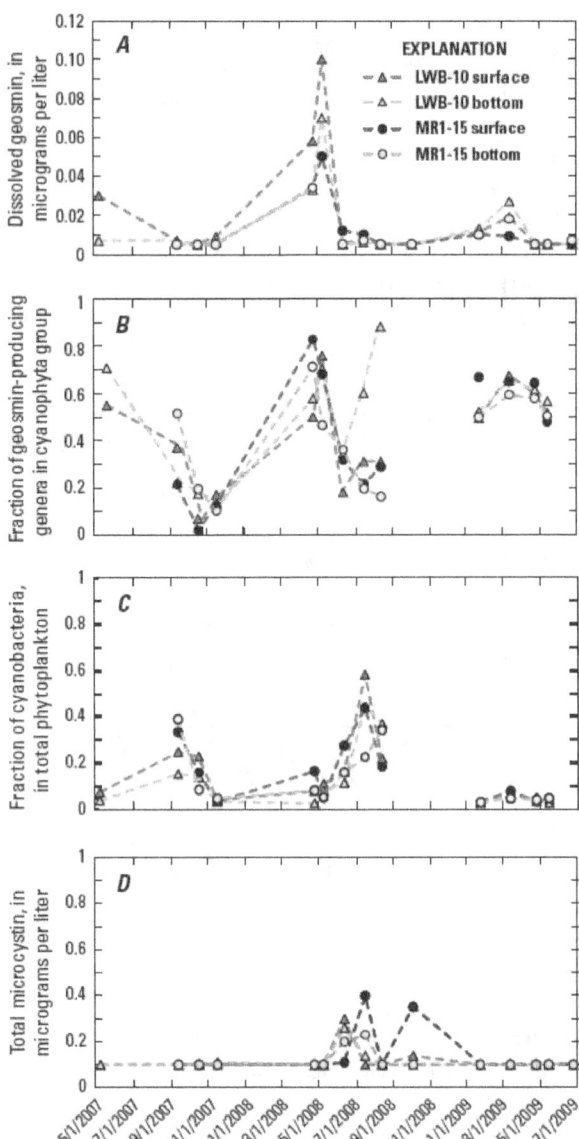

Figure 7. Temporal variation in the epilimnetic (surface) and hypolimnetic (bottom) concentrations of *(A)* dissolved geosmin, *(B)* geosmin-producing genera in cyanobacteria biovolume, *(C)* fraction of cyanobacteria in total algal biovolume, and *(D)* total microcystin at site LWB-10 in Lake William C. Bowen and at site MR1-15 in Municipal Resevoir #1, Spartanburg County, South Carolina, May 2007 to June 2009. Human taste threshold for geosmin is 0.010 microgram per liter. World Health Organization guidelines for microcystin are 1 microgram per liter (µg/L) for drinking water and 20 µg/L for recreational activities (Chorus and Bartram, 1999).

of 10 ng/L (0.010 µg/L). Maximum geosmin concentration at MR1-15 was only half (0.050 µg/L) the maximum concentration observed in Lake Bowen during the same spring 2008 period. Annual maximum geosmin concentrations tended to reoccur in the spring (April–May), but in 2009, the annual maximum geosmin concentrations occurred in March (fig. 7). Nonetheless, the March-to-May period of annual maximum geosmin concentrations was not concurrent with the period of annual maximum cyanobacterial biovolume (fig. 7). Cyanobacteria were present in both reservoirs during the peak geosmin period in the spring of 2008, but represented less than 20 percent of total phytoplankton biovolume (fig. 7). The peak geosmin period was concurrent with a peak in the fraction of known geosmin-producing genera in the cyanobacteria group (fig. 7). Microcystin rarely was detected. Maximum microcystin concentrations of 0.30 and 0.40 µg/L were observed in Lake Bowen and Reservoir #1, respectively, during the summer of 2008 immediately following the peak geosmin period (table 4; fig. 7). Microcystin concentrations were below concentrations of concern of 1 µg/L for drinking water or 20 µg/L for recreational activities (Chorus and Bartram, 1999). No statistical differences existed in geosmin, MIB, and microcystin concentrations and in cyanobacterial biovolume between Lake Bowen and Reservoir #1, among sites within each reservoir, and between depths at a site (table 4; fig. 6).

Because Lake Bowen releases directly into Reservoir #1, the potential of Lake Bowen as a significant source of geosmin to Reservoir #1 was evaluated. Predicted geosmin concentrations (nanograms per liter were used to simplify computations) in Reservoir #1 were computed from a simple flux model of geosmin concentrations near the Lake Bowen dam at site LWB-10 and daily inflow from Lake Bowen to Reservoir #1. Estimated geosmin flux was compared to observed geosmin concentrations at site MR1-12 (table 8). Site MR1-12 is the Reservoir #1 site closest to the Lake Bowen dam and, therefore, most influenced by Lake Bowen releases (fig. 1). Predicted geosmin concentrations from the simple flux model ranged from less than 0.1 to 3.4 ng/L, and observed geosmin concentrations at site MR1-12 ranged from less than 5 to 42 ng/L (table 8; fig. 8). It should be noted that all the estimated geosmin concentrations were below the analytical LRL of 5 ng/L and would have been reported as <5 ng/L for observed concentrations. Further comparison of the observed geosmin concentrations that were above the LRL to predicted geosmin concentrations in Reservoir #1 determined that predicted geosmin concentrations represented 2 to 14 percent of the observed geosmin concentrations at MR1-12 (table 8; fig. 8). Results of this simple model indicated that geosmin flux from Lake Bowen was probably not a major contributor of geosmin concentrations in Reservoir #1. Instead, in situ production and release of geosmin from cyanobacteria or actinomycetes likely were the predominant processes contributing to geosmin concentrations in Reservoir #1.

Table 8. Computed geosmin flux from site LWB-10 in Lake William C. Bowen to Municipal Reservoir #1 and estimated geosmin concentrations at site MR1-12 in Reservoir #1 attributed to the geosmin flux from Lake William C. Bowen, Spartanburg County, South Carolina, May 2007 to June 2009.

[ng/L, nanograms per liter; ft³/s, cubic feet per second; mg/d, milligrams per day; L, liters; <, less than]

Date of collection	Observed geosmin in source water (ng/L)			Total daily inflow to MR1 (ft³/s)	Total instantaneous geosmin flux to MR1 (mg/d)	Daily water volume in MR1 at MR1-12 (L)	Estimated geosmin at MR1-12	
	LWB-10-surface	LWB-10-bottom	MR1-12				Concentration (ng/L)	Percent of measured
05/07/07	30	7	12	42.7	3,130	1,837,433,160	1.7	14
04/28/08	58	33	42	24.2	3,440	1,691,695,260	2.0	5
05/12/08	100	70	30	24.2	5,930	1,732,199,040	3.4	11
06/09/08	5	5	<5	19.4	237	1,519,081,020	0.2	ND
07/21/08	8	6	5	12.9	190	1,796,550,840	0.1	2
08/18/08	<5	<5	<5	12.9	<158	1,651,948,560	<0.1	ND
10/14/08	5	<5	<5	12.9	<158	1,503,182,340	<0.1	ND
01/22/09	11	13	13	20.7	557	1,842,732,720	0.2	2
03/16/09	19	27	16	38.8	1800	1,768,538,880	1.0	6
04/27/09	<5	<5	<5	42.7	<522	1,795,036,680	<0.3	ND
05/11/09	<5	<5	<5	27.8	<340	1,859,388,480	<0.2	ND
06/22/09	<5	<5	<5	34.9	<427	1,845,003,960	<0.2	ND

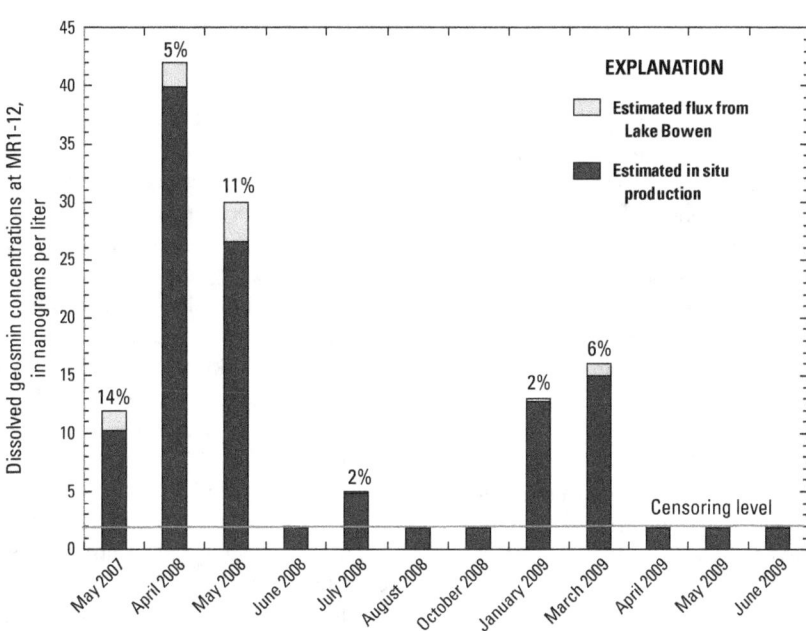

Figure 8. Dissolved geosmin concentrations at Municipal Reservoir #1 site MR1-12 estimated to be from geosmin flux from Lake William C. Bowen and from in situ production, Spartanburg County, South Carolina, May 2007 to June 2009.

Environmental Factors Influencing Geosmin Concentrations and Cyanobacterial Biovolumes

Further data analyses were limited to the surface sample data at site LWB-10 in Lake Bowen and at site MR1-15 in Reservoir #1. The low number of samples (less than 20) and the high number of geosmin concentrations below the LRL contributed to the limited number of significant correlations. In Lake Bowen at site LWB-10, elevated geosmin concentrations near the surface were correlated with deep (greater than 6 m) mixing zone (Z_m) depths (Kendall tau (τ) = 0.60), lower euphotic zone-to-mixing zone depth (Z_{eu}: Z_m) ratios ($\tau = -0.47$), and higher dissolved-oxygen concentrations ($\tau = 0.42$), which are environmental conditions indicative of unstratified conditions in Lake Bowen (table 9). In relation to phytoplankton community structure, elevated geosmin concentrations at LWB-10 were correlated to greater *Oscillatoria* biovolumes ($\tau = 0.60$). Elevated geosmin concentrations were correlated to reduced algal biomass (dry weight) and lower total phytoplankton biovolumes ($\tau = -0.61$ and -0.45, respectively; table 9). These correlations suggest that elevated geosmin concentrations tend to occur during periods of reduced algal biomass. Elevated geosmin concentrations also were correlated to higher chloride and silica concentrations ($\tau = 0.53$ and 0.51, respectively; table 9). Conversely, geosmin concentrations were not correlated to residence times, nutrient concentrations, chlorophyll *a* concentrations, actinomycetes concentrations, other known geosmin-producing genera of cyanobacteria, fraction of known geosmin-producing genera in the cyanobacteria group, or cyanobacterial biovolumes in Lake Bowen.

In Reservoir #1 at site MR1-15, elevated geosmin concentrations near the surface were correlated to greater dissolved-oxygen and chloride concentrations ($\tau = 0.55$ and 0.58, respectively; table 9). Elevated geosmin concentrations were correlated to greater chlorophyll *a*:pheophytin *a* ratios ($\tau = 0.80$), lower pheophytin *a* concentrations ($\tau = -0.61$) and to reduced algal biomass (dry weight) ($\tau = -0.53$; table 9). Elevated geosmin concentrations were correlated to greater fraction of geosmin-producing genera in the cyanobacteria group ($\tau = 0.56$); however, total cyanobacterial biovolumes were not correlated to geosmin concentrations. As was determined for Lake Bowen site LWB-10, 30-day residence times at site MR1-15 were not correlated to geosmin concentrations (table 9). Geosmin concentrations were not correlated to nutrient, chlorophyll *a*, or actinomycetes concentrations in surface samples at site MR1-15.

Site LWB-10 located in the lower region of Lake Bowen had geosmin concentrations correlated to the known geosmin-producing genus *Oscillatoria* within the cyanobacteria group but not to actinomycetes concentrations, suggesting that cyanobacteria were the probable source of geosmin. The corresponding site in Reservoir #1 (MR1-15) also had geosmin concentrations that correlated to known geosmin-producing genera as a fraction of the overall cyanobacteria group. Again, no significant correlation existed between geosmin and actinomycetes concentrations at this site.

In the deeper regions of the reservoirs (sites LWB-10, MR1-15), cyanobacterial biovolumes tended to be greatest during the summer when the reservoirs were stratified (tables 4, 5). In contrast to elevated geosmin concentrations in surface samples from Lake Bowen at site LWB-10, environmental factors indicative of stratified or stable water column conditions, including warmer water temperatures ($\tau = 0.86$), lower Z_m ($\tau = -0.84$), greater Z_{eu}:Z_m ratios ($\tau = 0.88$), lower dissolved-oxygen concentrations ($\tau = -0.52$), and maximum RTRM ($\tau = 0.91$), correlated strongly with elevated cyanobacterial biovolumes (table 9). Also, in contrast to elevated geosmin concentrations in surface samples, cyanobacterial biovolumes at site LWB-10 correlated with changes in nutrient levels in the reservoirs. At site LWB-10, elevated cyanobacterial volumes were correlated to lower nitrogen concentrations (as ammonia, nitrate plus nitrite, total organic nitrogen, and TN) and lower TN:TP ratios. Elevated cyanobacteria biovolume in both reservoirs correlated strongly to elevated hypolimnetic TP concentrations relative to the epilimnetic TP concentrations ($\tau = 0.55$) that were considered indicative of release of phosphorus from the sediment during anoxic conditions (table 9). Elevated cyanobacterial biovolumes in surface samples from Reservoir #1 at site MR1-15 had similar correlative relations to environmental factors as cyanobacterial biovolumes at site LWB-10, with the exception of no correlation to TN:TP ratios and a negative correlation to TP ($\tau = -0.51$) (table 9).

On the basis of 30-day moving window averages, elevated cyanobacterial biovolumes at sites LWB-10 and MR1-15 were correlated to longer residence times (as inflow, $\tau = 0.69$ and 0.50, respectively) and lower overflow volumes ($\tau = -0.61$ and -0.44, respectively), indicative of low-flow or drought conditions (table 9). Biovolumes of known geosmin- and toxin-producing cyanobacteria genera, including *Cylindrospermopsis*, *Planktolyngbya*, *Synechococcus*, *Synechocystis*, and *Aphanizomenon* (site LWB-10 only), correlated with greater cyanobacteria biovolumes and were the dominant taxa in the cyanobacteria group (table 9).

Production and release of geosmin often have been reported to be related to periods when cyanobacteria dominated the phytoplankton community and often produced species-specific blooms (Izaguirre and others, 1982; Smith, 1983; Downing and McCauley, 1992; Smith and others, 1995; Smith and Bennett, 1999; Jacoby and others, 2000; Downing and others, 2001; Pearl and others, 2001; Havens and others, 2003; Graham and others, 2004; Dzialowski and others, 2009). In turn, cyanobacterial dominance has been attributed to several environmental factors that allow the cyanobacteria to thrive more successfully than other phytoplankton groups, including decreased availability of nitrogen, increased phosphorus concentrations, reduced light availability, warmer water temperatures, greater water column stability, and longer residence times (Izaguirre and others, 1982; Smith, 1983; Downing and McCauley, 1992; Smith and others, 1995; Smith and Bennett, 1999; Jacoby and others, 2000; Downing and others, 2001; Pearl and others, 2001; Havens and others, 2003; Graham and others, 2004; Zaitlin and Watson, 2006; Dzialowski and others, 2009; Graham and Jones, 2009). In

Table 9. Kendall tau correlation coefficents for selected environmental variables, geosmin, and cyanobacterial biovolume for surface samples in Lake William C. Bowen at site LWB-10 and in Municipal Reservoir #1 at site MR1-15, Spartanburg County, South Carolina, May 2006 to June 2009.

[Bold highlighted tau indicates significant correlation at a probability value <0.05; mg/L, milligrams per liter; C, degrees Celsius; m, meters; Mgal/d, million gallons per day; col/100 mL, colonies per 100 milliliters; µg/L, micrograms per liter; µm³/mL, cubic micrometers per milliliter; %, percent; ND, no data]

Environmental factors	Geosmin (µg/L)		Cyanobacterial biovolume (µm³/mL)	
	LWB-10-1	MR1-15-1	LWB-10-1	MR1-15-1
Dissolved oxygen (mg/L)	**0.42**	**0.55**	**−0 52**	**−0.52**
Water temperature (C)	−0.38	−0.13	**0 86**	**0.73**
Transparency (m)	0.25	−0.13	0.41	−0.16
Euphotic zone depth (m)	0.33	0.21	**0 58**	0.19
Mixing zone depth (m)	**0.60**	0.25	**−0 84**	**−0.72**
Euphotic:mixing zone depth ratio	**−0.47**	−0.17	**0 88**	**0.47**
Maximum relative thermal resistance to mixing (unitless)	−0.39	−0.33	**0 91**	**0.84**
Hardness (mg/L)	−0.14	0.17	**0 53**	−0.09
Chloride (mg/L)	**0.53**	**0.58**	0 39	0.28
Silica (mg/L)	**0.51**	0.32	−0 29	−0.29
Ammonia (mg/L)	−0.01	−0.06	**−0.74**	**−0.68**
Nitrate plus nitrite (mg/L)	0.36	0.07	**−0 57**	**−0.55**
Dissolved inorganic nitrogen (mg/L)	0.37	−0.08	**−0 83**	**−0.64**
Total organic nitrogen (mg/L)	0.22	0.08	**−0 85**	**−0.63**
Total nitrogen (TN) (mg/L)	0.30	0.19	**−0.93**	**−0.64**
Total phosphorus (TP) (mg/L)	−0.14	−0.01	−0.37	**−0.51**
Hypolimnetic TP - Epilimnetic TP (mg/L)	−0.28	0.18	**0.55**	**0.71**
TN:TP ratio (unitless)	0.36	0.21	**−0.65**	−0.39
Chlorophyll a:TP ratio (unitless)	−0.03	−0.27	**−0.48**	−0.05
Average 30-day residence time—inflow (days)	0.02	0.14	**0.69**	**0.50**
Average 30-day residence time—outflow (days)	0.16	0.03	**0.59**	**0.47**
Average 30-day overflow (Mgal/d)	0.06	−0.03	**−0.61**	**−0.44**
Average 30-day flowthrough (Mgal/d)	−0.17	−0.14	0.38	0.27
Average 30-day overflow—Average 30-day flowthrough (Mgal/d)	0.06	0.00	**−0.53**	−0.39
Average 30-day outflow from Reservoir #1 (Mgal/d)	−0.11	−0.04	**−0.44**	−0.37
Actinomycetes (col/100 mL)	−0.33	−0.14	−0.08	−0.39
Biomass:chlorophyll a ratio (unitless)	−0.36	−0.05	**0.46**	0.31
Algal biomass (ash weight) (mg/L)	**−0.58**	**−0.54**	0 11	0.35
Algal biomass (dry weight) (mg/L)	**−0.61**	**−0.53**	0.01	0.38
Chlorophyll a (µg/L)	0.01	−0.27	**−0.65**	−0.17
Pheophytin a (µg/L)	−0.25	**−0.61**	**−0 55**	−0.08
Chlorophyll:pheophytin ratio (unitless)	0.38	**0.80**	0.46	0.07
Dissolved iron (µg/L)	−0.32	−0.25	**−0 54**	**−0.51**
Dissolved manganese (µg/L)	0.02	−0.02	**−0.78**	**−0.71**
Total microcystin (µg/L)	−0.32	0.14	0 31	**0.52**
Dissolved geosmin (µg/L)	ND	ND	−0 29	−0.11
Dissolved organic carbon (mg/L)	−0.19	−0.17	0 30	0.00
Ultraviolet absorbance at 254 nanometers (cm⁻¹)	−0.39	−0.31	**−0 52**	−0.14
Total phytoplankton biovolume (µm³/mL)	**−0.45**	−0.27	0 29	**0.43**
Total cyanobacteria biovolume (µm³/mL)	−0.29	−0.11	ND	ND
Fraction of cyanobacteria in total phytoplankton (%)	−0.10	0.02	**0 89**	**0.94**
Fraction of geosmin-producing genera in cyanobacteria group (%)	0.34	**0.56**	−0.02	−0.24
Anabaena biovolume (µm³/mL)	−0.29	−0.16	**0 52**	**0.56**
Aphanizomenon biovolume (µm³/mL)	−0.24	−0.00	**0.63**	0.41
Cylindrospermopsis biovolume (µm³/mL)	−0.39	−0.37	**0.69**	**0.51**
Oscillatoria biovolume (µm³/mL)	**0.60**	0.11	−0 10	−0.06
Phormidium biovolume (µm³/mL)	**−0.46**	ND	−0 15	ND
Planktolyngbya biovolume (µm³/mL)	−0.23	−0.23	**0.65**	**0.70**
Pseudoanabaena biovolume (µm³/mL)	0.20	0.17	−0 20	0.02
Microcystis biovolume (µm³/mL)	−0.39	**−0.42**	0 34	0.27
Synechococcus biovolume (µm³/mL)	−0.04	0.29	**0.68**	**0.75**
Synechocystis biovolume (µm³/mL)	**−0.45**	−0.24	0 59	**0.83**

Lake Bowen and Reservoir #1, cyanobacterial biovolumes were related to many of these reported environmental factors as well as increased total phosphorus in the hypolimnion attributed to sediment release during anoxic conditions (internal phosphorus cycling). Elevated dissolved geosmin concentrations, however, were not related to increased cyanobacterial biovolumes, and cyanobacteria rarely dominated the total phytoplankton community. Therefore, another mechanism was needed to explain the increased production and release of geosmin in April and May of 2008 for Lake Bowen and Reservoir #1.

One plausible explanation of elevated geosmin concentrations in 2008 is related to a pattern of greater transparencies observed in both reservoirs (more pronounced in Lake Bowen) concurrent with annual maximum dissolved geosmin concentrations in the spring. In fact, during the spring of 2008, maximum transparencies (3.7 m in Lake Bowen and 2.0 m in Reservoir #1) coincided with the period of peak dissolved geosmin production when maximum geosmin concentrations were 100 ng/L in Lake Bowen and 50 ng/L in Reservoir #1. That springtime pattern of greater transparency is consistent with a clear water phase usually attributed to heavy zooplankton grazing of the phytoplankton community and has been reported commonly in mesotrophic reservoirs (Durrer and others, 1999; Scheffer, 2004). Because zooplankton data were not available, the occurrence of this process could not be confirmed directly. Nonetheless, the relation between elevated dissolved geosmin concentrations and environmental factors other than deeper transparencies also suggested zooplankton grazing could be a mechanism for the direct or indirect release of geosmin from cyanobacteria into the dissolved phase. In both reservoirs, elevated dissolved geosmin concentrations were correlated to environmental factors not only indicative of greater light penetration (greater euphotic zone depths) but also reduced algal biomass and total phytoplankton biovolumes. Geosmin release from cyanobacterial cells has been associated with cellular senescence or cell lysis (Rashash and others, 1996) and with periods of environmental stress (Paerl and Millie, 1996; Watson, 2003), all of which are consistent with heavy zooplankton grazing. While cyanobacteria generally are not considered favorable food for zooplankton, a shift in the zooplankton community toward less selective predation is probable because of the increased competition for prey that would accompany a heavy zooplankton grazing event (Scheffer, 2004; Sarnelle and Wilson, 2005; Hansson and others, 2007). Another indication of heavy zooplankton grazing was the coincidence of the lowest total phytoplankton and cyanobacterial biovolumes near the surface (1-m depth) during the period of maximum dissolved geosmin concentrations in the spring of 2008 (fig. 7).

Predator-driven natural selection of cyanobacterial genera with chemical-defense capabilities could further contribute to elevated dissolved geosmin concentrations. During the period of maximum geosmin concentrations, three genera with known geosmin-producing strains in the cyanobacteria group (*Oscillatoria, Aphanizomenon,* and *Synechococcus*) were the dominant cyanobacteria taxa in both reservoirs. Increasing grazing pressure with decreasing prey alternatives would be expected to trigger chemical defenses in this surviving population. While grazing of *Oscillatoria* and *Aphanizomenon* by zooplankton is not well documented, the genus *Synechococcus* consists of picoplankton that are important contributors to pelagic freshwater ecosystems and have been shown to be grazed by zooplankton (Fahnenstiel and others, 1991; Callieri, and others, 2004).

Multiple Logistic Regression Model of Geosmin Concentrations

Three multiple logistic regression models (MLogModel) were developed that estimated the occurrence of geosmin concentrations above the human detection threshold of 10 ng/L (0.010 µg/L) on the basis of the multivariate and Kendall tau correlation analysis. Even though significant correlation existed between geosmin and several phytoplankton taxonomic variables (for example, total phytoplankton and *Oscillatoria* biovolumes), explanatory variables used as input for the MLogModels were limited to more easily or quickly measured chemical and hydrodynamic parameters. The three models were labeled MLogRModel1—developed from data at the surface depth (1 m) at LWB-10, MLogRModel2—developed from data at the surface depth (1 m) at MR1-15, and MLogRModel3—developed from data at the surface depth (1 m) at MR1-15 and total geosmin concentrations in the raw water at R.B. Simms Water Treatment Plant. Although many of the same environmental factors were used as explanatory variables in initial model runs, the final explanatory variables that provided the best fit model varied among the three sites (table 10).

The best fit model for MLogModel1 for LWB-10 was the following:

$$\text{Logit } P = -4.691 + (2.184 * [Z_m]) - (24.419 * \sqrt{[TN]}) + (0.0351 * [\text{Overflow} - \text{Flowthrough}]), \quad (2)$$

where Z_m is mixing zone depth in meters, $\sqrt{TN}$ is the square root of the total nitrogen concentration in milligrams per liter, and Overflow – Flowthrough is the 30-day prior moving window average of overflow minus the 30-day prior moving-window average of overflow at Lake Bowen dam, in million gallons per day. At site LWB-10, the likelihood of dissolved geosmin concentrations exceeding the human detection threshold was estimated by greater mixing zone depths and differences in the 30-day prior moving-window averages of overflow and flowthrough at Lake Bowen dam and by lower total nitrogen concentrations as described by MLogModel1. Of the three explanatory variables, only mixing zone depth was correlated significantly to geosmin concentrations in the exploratory data analysis at LWB-10 (table 10). The MLogModel1 correctly estimated the likelihood of geosmin concentrations exceeding the human detection threshold 83 percent of the time and not exceeding the human detection threshold 100 percent of the time, resulting in an overall sensitivity of 94 percent (table 10).

Table 10. Classification tables for the multiple logistic regression models developed to estimate the likelihood of geosmin concentrations exceeding the human detection threshold of 10 nanograms per liter in Lake William C. Bowen at site LWB-10 (MLogModel1), in Municipal Reservoir #1 at site MR1-15 (MLogModel2), and at R.B. Simms Water Treatment Plant (MLogModel3).

Classification table	Predicted reference	Predicted positive	Totals	Diagnostic	Percent	Hosmer-Lemshow (p-value)	Pearson Chi-Squared (p-value)	Likelihood Ratio Test (p-value)
MLogModel1 for LWB-10								
Actual reference responses	11	0	11	Specificity	100	9.07 (0.336)	15.49 (0.216)	12.12 (0.007)
Actual positive responses	1	5	6	Sensitivity	83			
Totals	12	5	17	Overall	94			
MLogRModel2 for MR1-15								
Actual reference responses	9	1	10	Specificity	90	2.014 (0.981)	5.09 (0.827)	8.593 (0.014)
Actual positive responses	1	2	3	Sensitivity	67			
Totals	10	3	13	Overall	85			
MLogModel3 for R.B. Simms								
Actual reference responses	3	1	4	Specificity	75	0.842 (0.999)	6.309 (0.789)	9.928 (0.007)
Actual positive responses	1	9	10	Sensitivity	90			
Totals	4	10	14	Overall	86			

The best fit model for MLogModel2 for MR1-15 was the following:

$$\text{Logit P} = -21.251 - (4.949 * \text{Log10[MR1 Outflow]}) + (4.343 * [Z_{eu}]), \tag{3}$$

where Log10[MR1 Outflow] is the logarithm of the 30-day prior moving window average of outflow from Reservoir #1, in million gallons per day, and Z_{eu} is the euphotic zone depth, in meters. Outflow from Reservoir #1 was computed as a 30-day prior moving-window average of differences in streamflow between the Pacolet River USGS streamgaging station (02155500) and North Pacolet River USGS streamgaging station (02154500). Euphotic zone depth, in meters, extended from the surface of the water downward to a depth where light intensity fell to 1 percent of that at the surface. At site MR1-15, greater euphotic zone depth and reduced outflow from Reservoir #1 provided the best estimation of dissolved geosmin concentrations above the human detection threshold in MLogModel2. Of the two explanatory variables, only euphotic zone depth was correlated significantly to geosmin concentrations (table 9). The MLogModel2 was less sensitive than MLogModel1. When applied to the dataset at site MR1-15, MLogModel2 correctly estimated the likelihood of geosmin concentrations exceeding the human detection threshold 67 percent of the time and not exceeding the human detection threshold 90 percent of the time, resulting in an overall sensitivity of 85 percent (table 10). The reduced sensitivity of the MLogModel2 for site MR1-15 relative to the other two models was attributed, in part, to the fact that the site had a much lower number of observed positive responses (geosmin concentrations above the human detection threshold) than the other sites, which decreased the ability of the regression model to accurately estimate that response.

The best fit model for MLogModel3 for R.B. Simms WTP was the following:

$$\text{Logit P} = -5.735 - (48.364 * [\text{DIN}]) + (9.724 * \text{Log10[MR1 Outflow]}), \tag{4}$$

where DIN is the dissolved inorganic nitrogen (dissolved ammonia and nitrate plus nitrite) concentration, in milligrams per liter, and Log10[MR1 Outflow] is the logarithm of the 30-day prior moving-window average of outflow from Reservoir #1, in million gallons per day. At the R.B. Simms Water Treatment Plant, the likelihood of total geosmin concentrations in the raw water exceeding the human detection threshold was estimated by greater outflow from Reservoir #1 and lower concentrations of dissolved inorganic nitrogen (combined dissolved ammonia and nitrate plus nitrite) in MLogModel3. MLogModel3 correctly estimated the likelihood of total geosmin concentrations exceeding the human detection threshold 90 percent of the time and not exceeding the human detection threshold 75 percent of the time, resulting in an overall sensitivity of 86 percent (table 10).

Overall, MLogModel1 and 3 indicated greater likelihood for both dissolved and total geosmin to exceed the human detection threshold during periods of lower nitrogen concentrations (total in Lake Bowen and dissolved inorganic in Reservoir #1) and greater water movement in the reservoir (greater overflow relative to flowthrough in Lake Bowen and greater outflow in Reservoir #1). Conversely, MLogModel2 indicated a greater likelihood of threshold exceedences by dissolved geosmin concentrations during periods of reduced outflow from Reservoir #1 and greater light penetration. It also should be noted that the calibration dataset for the logistic model had a small sample size (less than 20 samples) and was collected during a hydrologic period of extremely low-flow to average conditions. The small sample size and extreme hydrologic conditions may limit the applicability of these models for above-average flow conditions and, especially, for high-flow conditions.

Summary

The occurrence of dissolved geosmin was studied in two reservoirs in Spartanburg County, South Carolina, from May 2006 to June 2009. Lake Bowen and Reservoir #1 are relatively shallow, meso-eutrophic, warm monomictic, cascading impoundments on the South Pacolet River. Overall, water-quality conditions and phytoplankton community assemblages were similar between the two reservoirs but differed seasonally. Median dissolved geosmin concentrations in the reservoirs ranged from 0.004 to 0.006 microgram per liter (μg/L). Annual maximum dissolved geosmin concentrations tended to occur between March and May. In this study, peak dissolved geosmin production occurred in April and May 2008, ranging from 0.050 to 0.100 μg/L at the deeper reservoir sites. Although Lake Bowen discharges directly into Reservoir #1, the geosmin flux from Lake Bowen contributed minimally to the geosmin concentration in Reservoir #1. Instead, in situ production of geosmin by cyanobacteria was the most probable source of elevated geosmin concentrations in Reservoir #1 and Lake Bowen.

In Lake Bowen, elevated geosmin concentrations near the surface were correlated to environmental conditions indicative of unstratified conditions (higher dissolved-oxygen concentrations and greater Z_m). In relation to phytoplankton community structure, elevated geosmin concentrations were correlated to greater *Oscillatoria* biovolumes, a genus of cyanobacteria with known geosmin-producing species. Elevated geosmin concentrations were correlated to reduced algal biomass and lower total phytoplankton biovolumes. In Reservoir #1, elevated geosmin concentrations near the surface were correlated to greater dissolved-oxygen concentrations and to reduced algal biomass. Rather than a specific genus of cyanobacteria, elevated geosmin concentrations in Reservoir #1 were correlated to a greater fraction of geosmin-producing genera in the cyanobacteria group. However, total cyanobacterial biovolumes were not correlated to geosmin concentrations.

In contrast to elevated geosmin concentrations in surface samples from Lake Bowen and Reservoir #1, environmental factors indicative of stratified or stable water column conditions, including warmer water temperatures, lower Z_m, greater $Z_{eu}:Z_m$ ratios, lower dissolved-oxygen concentrations, and maximum RTRM, correlated strongly with elevated cyanobacterial biovolumes. Also, in contrast to elevated geosmin concentrations in surface sample, elevated cyanobacterial biovolumes correlated with changes in nutrient levels in the reservoirs, including lower nitrogen concentrations (as ammonia, nitrate plus nitrite, total organic nitrogen, and TN) and elevated hypolimnetic TP concentrations relative to the epilimnetic TP concentrations (considered indicative of release of phosphorus from the sediment during anoxic conditions). In both reservoirs, elevated cyanobacterial biovolumes were correlated to longer residence times and lower overflow volumes, indicative of low-flow or drought conditions. A greater fraction of cyanobacteria in the total phytoplankton community and biovolumes of known geosmin- and toxin-producing cyanobacteria genera, including *Cylindrospermopsis, Planktolyngbya, Synechococcus, Synechocystis,* and *Aphanizomenon,* correlated with the greater cyanobacteria biovolumes and were the dominant taxa in the cyanobacteria group.

In Lake Bowen and Reservoir #1, elevated cyanobacterial biovolumes were related to many environmental factors that have been previously reported to enhance cyanobacterial dominance of the phytoplankton community, including decreased availability of nitrogen, increased phosphorus concentrations, reduced light availability, warmer water temperatures, greater water column stability, and longer residence times. However, unlike previous reports, elevated dissolved geosmin concentrations were not related to increased cyanobacterial biovolumes and cyanobacteria rarely dominated the total phytoplankton community. Therefore, another mechanism was needed that could explain the increased production and release of geosmin in April and May of 2008.

One plausible explanation of elevated geosmin concentrations in 2008 is related to a pattern of greater transparencies observed in both reservoirs (more pronounced in Lake Bowen) concurrent with annual maximum dissolved geosmin concentrations in the spring. In fact, during spring 2008, maximum transparencies (3.7 meters in Lake Bowen and 2.0 meters in Reservoir #1) coincided with the period of peak dissolved geosmin production when maximum geosmin concentrations were 100 ng/L in Lake Bowen and 50 ng/L in Reservoir #1. That springtime pattern of greater transparency is consistent with a "clear water phase" usually attributed to heavy zooplankton grazing of the phytoplankton community as has been reported commonly in mesotrophic reservoirs. Because zooplankton data were not available, the actual occurrence of this process could not be confirmed.

Logistic regression models indicated geosmin concentrations had the greatest probability (83 percent model sensitivity) of exceeding 10 ng/L during periods of greater overflow (higher water levels in Lake Bowen) relative to flowthrough releases at the dam, lower total nitrogen, and unstratified conditions (greater mixing zone depths) at site LWB-10. Similarly, raw water geosmin concentrations above 10 ng/L at the R.B. Simms Water Treatment Plant were probable (86 percent model sensitivity) during periods of greater outflow at the Reservoir #1 dam and lower dissolved inorganic nitrogen concentrations at MR1-15. Conversely, in the source water in Reservoir #1 at site MR1-15, geosmin concentrations above 10 ng/L were more probable (only a 67 percent model sensitivity) during periods of lower outflow but greater light penetration (euphotic zone depth, that correlated to transparency). Fewer periods of geosmin concentrations exceeding 10 ng/L (only 3 compared to 6 and 14 at the other two sites) could have produced the reduced sensitivity, poorer fit, and apparent inverse relations of elevated geosmin concentrations to hydrodynamic conditions relative to LWB-10 and the raw water. It also should be noted that the calibration dataset for the logistic model had a small sample size (less than 20 samples) and was collected during a hydrologic period of extremely low-flow to average conditions. The small sample size and extreme hydrologic conditions potentially may limit the applicability of these models for above-average flow conditions and, especially, for high-flow conditions.

Spartanburg Water can include the logistic regression model parameters in their ongoing geosmin monitoring program in Lake Bowen and Reservoir #1. Additional information including the potential link between increased cyanobacterial biovolumes and internal phosphorus cycling identified a need for further investigation. Ultimately, Spartanburg Water can integrate findings from this investigation into watershed management plans designed to reduce the occurrence of cyanobacteria and associated nuisance compounds.

References

American Public Health Association, American Water Works Association, and Water Environment Federation, 1995a, Standard methods for the examination of water and wastewater (19th ed.), UV-absorbing organic constituents: American Public Health Association, part 5910, p. 5-60 to 5-62.

American Public Health Association, American Water Works Association, and Water Environment Federation, 1995b, Standard methods for the examination of water and wastewater (19th ed.), Determination of biomass (standing crop): American Public Health Association, part 10221, p. 10–25.

American Public Health Association, American Water Works Association, and Water Environment Federation, 1998, Standard methods for the examination of water and wastewater (20th ed.): Washington, D.C., American Public Health Association, p. 3-37 to 3-43.

American Public Health Association, American Water Works Association, and Water Environment Federation, 2005, Standard methods for the examination of water and wastewater (21st ed.), Detection of actinomycetes: American Public Health Association, part 9250, p. 9-109 to 9-111.

Arar, E.J., and Collins, G.B., 1997, U.S. Environmental Protection Agency Method 445.0, In vitro determination of chlorophyll *a* and pheophytin *a* in marine and freshwater algae by fluorescence, Revision 1.2: Cincinnati, Ohio, U.S. Environmental Protection Agency, National Exposure Research Laboratory, Office of Research and Development, 22 p.

Brenton, R.W., and Arnett, T.L., 1993, Methods of analysis by the U.S. Geological Survey National Water Quality Laboratory—Determination of dissolved organic carbon by uv-promoted persulfate oxidation and infrared spectrometry: U.S. Geological Survey Open-File Report 92–480, 12 p.

Brookes, J.D., Daly, R., Regel, R.H., Burch, M., Ho, L., Newcombe, G., Hoefel, D., Saint, C., Meyne, T., Buford, M., Smith, M., Shaw, G., Guo, P.P., Lewis, D., and Hipsey, M., 2008, Reservoir management strategies for control and degradation of algal toxins: American Water Works Association, 241 p.

Burkholder, J.M., 1992, Phytoplankton and episodic suspended sediment loading—Phosphate partitioning and mechanisms for survival: Limnology and Oceanography, v. 37, no. 5, p. 974–988.

Callieri, C., Balseiro, E., Bertoni, R., and Modenutti, B., 2004, Picocyanobacterial photosynthetic efficiency under *Daphnia* grazing pressure: Journal of Plankton Research, v. 26, no. 12, p. 1471–1477.

Carmichael, W.W., 1994, The toxins of cyanobacteria: Scientific American, v. 270, p. 78–86.

Carpenter, S.R., Elser, M.M., and Elser, J.J., 1986, Chlorophyll production, degradation, and sedimentation—Implications for paleolimnology: Limnology and Oceanography, v. 31, p. 112–124.

Cartaxana, P., Jesus, B., and Brotas, V., 2003, Pheophorbide and pheophytin *a*-like pigments as useful markers for intertidal microphytobenthos grazing by *Hydrobia ulvae*: Estuarine, Coastal and Shelf Science, v. 58, p. 299–303.

Childress, C.J.O., Foreman, W.T., Conner, B.F., and Maloney, T.J., 1999, New reporting procedures based on long-term method detection levels and some considerations for interpretations of water-quality data provided by the U.S. Geological Survey National Water Quality Laboratory: U.S. Geological Survey Open-File Report 99-193, 19 p. (Also available at *http://water.usgs.gov/owq/OFR_99-193/*.)

Chorus, I., and Bartram, J., 1999, Toxic cyanobacteria in water—A guide to their public health consequences, monitoring and management: London, E & FN Spon/Chapman & Hall, 416 p.

Christensen, V.G., Graham, J.L., Milligan, C.R., Pope, L.M., and Ziegler, A.C., 2006, Water quality and relation to taste-and-odor compounds in North Fork Ninnescah River and Cheney Reservoir, south-central Kansas, 1997–2003: U.S. Geological Survey Scientific Investigations Report 2006–5095, 43 p.

Clarke, K.R., and Gorley, R.N., 2006, PRIMER (version 6)—Users manual/Tutorial: Plymouth, U.K., PRIMER-E, 192 p.

Clarke, K.R., and Warwick, R.M., 2001, Change in marine communities—An approach to statistical analysis and interpretation (2d ed.): Plymouth, U.K., PRIMER-E, 172 p.

Cooke, G.D., Welch, E.B., Peterson, S.A., and Nichols, S.A., 2005, Restoration and management of lakes and reservoirs: Boca Raton, Florida, Taylor and Francis Group, 591 p.

Cuker, B.E., Gama, P.T., and Burkholder, J.M., 1990, Type of suspended clay influences lake productivity and phytoplankton community response to phosphorus loading: Limnology and Oceanography, v. 35, no. 4, p. 830–839.

Dokulil, M.T., and Teubner, K., 2000, Cyanobacterial dominance in lakes: Hydrobiologica, v. 438, p. 1–12.

Downing, J.A., and McCauley, E., 1992, The nitrogen: phosphorus relationship in lakes: Limnology and Oceanography, v. 37, no. 5, p. 936–945.

Downing, J.A., Watson, S.B., and McCauley, E., 2001, Predicting cyanobacteria dominance in lakes: Canadian Journal of Fishery and Aquatic Sciences, v. 58, p. 1905–1908.

Durrer, M., Zimmermann, U., and Jüttner, F., 1999, Dissolved and particle-bound geosmin in a mesotrophic lake (Lake Zurich)—Spatial and seasonal distribution and the effect of grazers: Water Research, v. 33, no. 17, p. 3628–3636.

Dzialowski, A.R., Smith, V.H., Huggins, D.G., deNoyelles, F., Lim, N.-C., Baker, D.S., and Beury, J.H., 2009, Development of predictive models for geosmin-related taste and odor in Kansas, USA, drinking water reservoirs: Water Research, v. 43, p. 2829–2840.

Fahnenstiel, G.L., Carrick, H.J., and Iturriaga, R., 1991, Physiological characteristics and food-web dynamics of *Synechococcus* in Lakes Huron and Michigan: Limnology and Oceanography, v. 36, no. 2, p. 219–234.

Fishman, M.J., ed., 1993, Methods of analysis by the U.S. Geological Survey National Water Quality Laboratory—Determination of inorganic and organic constituents in water and fluvial sediments: U.S. Geological Survey Open-File Report 93–125, 217 p.

Fishman, M.J., and Friedman, L.C., eds., 1989, Methods for determination of inorganic substances in water and fluvial sediments: U.S. Geological Survey Techniques of Water-Resources Investigations, book 5, chap. A1, 545 p.

Forsberg, C., and Ryding, S., 1980, Eutrophication parameters and trophic state indices in 30 Swedish waste-receiving lakes: Archiv für Hydrobiologie, v. 89, p. 189–207.

Fry, J.A., Coan, M.J., Homer, C.G., Meyer, D.K., and Wickham, J.D., 2009, Completion of the National Land Cover Database (NLCD) 1992–2001 Land Cover Change Retrofit product: U.S. Geological Survey Open-File Report 2008–1379, 18 p.

Graham, J.L., Jones, J.R., Jones, S.B., Downing, J.A., and Clevenger, T.E., 2004, Environmental factors influencing microcystin distribution and concentration in the Midwestern United States: Water Research, no. 38, p. 4395–4404.

Graham, J.L., and Jones, J.R., 2009, Microcystin in Missouri reservoirs: Lake and Reservoir Management, v. 25, p. 253–263.

Graham, J.L., Loftin, K.A., and Kamman, N., 2009, Monitoring recreational freshwaters: LakeLine, v. 29, no. 2, p. 18–24.

Graham, J.L., Loftin, K.A., Meyer, M.T., and Ziegler, A.C., 2010, Cyanotoxin mixtures and taste-and-odor compounds in cyanobacterial blooms from the Midwestern United States: Environmental Science and Technology, v. 44, no. 19, p. 7361–7373.

Graham, J.L., Loftin, K.A., Ziegler, A.C., and Meyer, M.T., 2008, Guidelines for design and sampling for cyanobacterial toxin and taste-and-odor studies in lakes and reservoirs: U.S. Geological Survey Scientific Investigations Report 2008–5038, 39 p.

Hansson, L.-A., Gustafsson, S., Rengefors, K., and Bomark, L., 2007, Cyanobacterial chemical warfare affects zooplankton community composition: Freshwater Biology, v. 52, p. 1290–1301.

Havens, K.E., James, R.T., East, T.L., and Smith, V.H., 2003, N:P ratios, light limitation, and cyanobacteria dominance in a subtropical lake impacted by non-point source nutrient pollution: Environmental Pollution, v. 122, p. 379–390.

Helsel, D.R., 2005, Nondetects and data analysis—Statistics for censored environmental data: Wiley, New York, 250 p.

Helsel, D.R., and Hirsch, R.M., 1992, Statistical methods in water resources: Amsterdam, The Netherlands, Elsevier Science Publishers, 522 p.

Homer, C., Huang, C., Yang, L., Wylie, B., and Coan, M., 2004, Development of a 2001 National Landcover Database for the United States—Photogrammetric Engineering and Remote Sensing, v. 70, no. 7, p. 829–840.

Ikawa, M., Sasner, J.J., and Haney, J.F., 2001, Activity of cyanobacterial and algal odor compounds fournd in lake waters on green alga *Chlorella pyrenoidosa* growth: Hydrobiologia, v. 443, p. 19–22.

Izaguirre, G., Hwang, C.J., Krasner, S.W., and McGuire, M.J., 1982, Geosmin and 2-methylisoborneol from cyanobacteria in three water supply systems: Applied and Environmental Microbiology, v. 43, no. 3, p. 708–714.

Jacoby, J.M., Collier, D.C., Welch, E.B., Hardy, F.J., and Crayton, M., 2000, Environmental factors associated with a toxic bloom of *Microcystis aeruginosa*: Canadian Journal of Fishery and Aquatic Sciences, v. 57, p. 231–240.

Journey, C.A., and Abrahamsen, T.A., 2008, Limnological conditions in Lake William C. Bowen and Municipal Reservoir #1, Spartanburg County, South Carolina, August to September 2005, May 2006, and October 2006: U.S. Geological Survey Open-File Report 2008–1268, 96 p. (Also available at *http://pubs.usgs.gov/of/2008/1268/*.)

Journey, C.A., Caldwell, A.W., Feaster, T.D., Petkewich, M.D., and Bradley, P.M., 2011, Concentrations, loads, and yields of nutrient and suspended sediment in the South Pacolet, North Pacolet, and Pacolet Rivers, northern South Carolina and southwestern North Carolina, October 2005 to September 2009: U.S. Geological Survey Scientific Investigations Report 2010–5252, 79 p.

Jüttner, F., and Watson, S.B., 2007, Biochemical and ecological control of geosmin and 2-methylisoborneol in source waters: Applied and Environmental Microbiology, v. 73, no. 14, p. 4395–4406.

Kronberg, N., and Purvis, J.C., 1959, Climates of the States—South Carolina, *in* Climatography of the States: U.S. Department of Commerce, Washington, D.C.

Nagle, D.D., Campbell, B.G., and Lowery, M.A., 2009, Bathymetry of Lake William C. Bowen and Municipal Reservoir #1, Spartanburg County, South Carolina, 2008: U.S. Geological Survey Scientific Investigations Map 3076, 1 sheet.

National Climatic Data Center, 2004, Climatography of the United States No. 20—1971–2000 for Greenville-Spartanburg Airport Station near Greer, S.C., accessed July 26, 2010, at *http://www.ncdc.noaa.gov/oa/climate/normals/usnormals.html*.

Nürnberg, G.K., 1996, Trophic state of clear and colored, soft- and hardwater lakes with special consideration of nutrients, anoxia, phytoplankton and fish: Lake and Reservoir Management, v. 12, p. 432–447.

Paerl, H.W., 1988, Nuisance phytoplankton blooms in coastal, estuarine, and inland waters: Limnology and Oceanography, v. 33, p. 823–847.

Paerl, H.W., Fulton, R.S., Moisander, P.H., and Dyble, J., 2001, Harmful freshwater algal blooms, with an emphasis on cyanobacteria: The Scientific World, v. 1, p. 76–113.

Paerl, H.W., and Millie, D.F., 1996, Physiological ecology of toxic aquatic cyanobacteria: Phycologia, v. 35, no. 6S, p. 160–167.

Patton, C.J., and Kryskalla, J.R., 2003, Methods of analysis of the U.S. Geological Survey National Water Quality Laboratory—Evaluation of alkaline persulfate digestion as an alternative to Kjeldahl digestion for determination of total and dissolved nitrogen and phosphorus in water: U.S. Geological Survey Water-Resources Investigations Report 03–4174, 33 p.

Peters, A., Köster, O., Schildknecht, A., and von Gunten, U., 2009, Occurrence of dissolved and particle-bound taste and odor compounds in Swiss lake waters: Water Research, v. 43, p. 2191–2200.

Pilotto, L.S., Kliewer, E.V., Davies, R.D., Burch, M.D., Attewell, R.G., 1999, Cyanobacterial (blue-green algae) contamination in drinking water and perinatal outcomes: Australian and New Zealand Journal of Public Health, v. 23, p. 154–158.

Purvis, J.C., Tyler, W., and Sidlow, S.F., 1990, Climate Report G-5—General Characteristics of South Carolina's Climate: Columbia, South Carolina Department of Natural Resources Water Resources Division (formerly South Carolina Water Resources Commission).

Rashash, D., Hoehn, R., Dietrich, A., Grizzard, T., and Parker, B., 1996, Identification and control of odorous algal metabolites: American Water Works Association (AWWA) Research Foundation and AWWA, 242 p.

Sarnelle, O., and Wilson, A.E., 2005, Local adaptation of *Daphnia pulicaria* to toxic cyanobacteria: Limnology and Oceanography, v. 50, p. 1565–1570.

Scheffer, M., 2004, Ecology of shallow lakes: The Netherlands, Kluwer Academic Publishers, 357 p.

Schuman, F.R., and Lorenzen, C.J., 1975, Quantitative degradation of chlorophyll by a marine herbivore: Limnology and Oceanography, v. 20, no. 4, p. 580–586.

Sklenar, K.S., and Horne, A.J., 1999, Effect of the cyanobacterial metabolite geosmin on growth of a green alga: Water Science and Technology, v. 40, p. 225–228.

Smith, V.H., 1983, Low nitrogen to phosphorus ratios favor dominance by blue-green algae in lake phytoplankton: Science, v. 221, p. 669–671.

Smith, V.H., and Bennett, S.J., 1999, Nitrogen:phosphorus supply ratios and phytoplankton community structure: Archiv für Hydrobiologie, v. 146, p. 37–53.

Smith, V.H., Bierman, V.J., Jones, B.L., and Havens, K.E., 1995, Historical trends in the Lake Okeechobee ecosystem IV—Nitrogen:phosphorus ratios, cyanobacterial dominance, and nitrogen fixation potential: Archiv für Hydrobiologie, Monographische Beitrage, no. 107, p. 71–88.

Smith, V.H., Sieber-Denlinger, J., deNoyelles, F., Jr., Campbell, S., Pan, S., Randke, S.J., Blain, G.T., and Strasser, V.A., 2002, Managing taste and odor problems in a eutrophic drinking water reservoir: Lake and Reservoir Management, v. 18, no. 4, p. 319–323.

South Carolina Department of Health and Environmental Control, 2006, Water classifications and standards: South Carolina Department of Health and Environmental Control, Code of Regulations, State Register, Regulation 61–68, accessed January 3, 2007, at *http://www.scdhec.net/ environment/water/regs/r61-68.doc*.

Spooner, N., Keely, B.J., and Maxwell, J.R., 1994, Biologically mediated defunctionalization of chlorophyll in the aquatic environment—I. Senescence/decay of the diatom *Phaeodactylum tricornutum*: Organic Geochemistry, v. 21, p. 509–516.

Sterner, R.W., 1989, Resource competition during seasonal succession toward dominance by cyanobacteria: Ecology, v. 70, p. 229–245.

Suffet, I.H., Corado, A., Chou, D., Butterworth, S., and MacGuire, M.J., 1996, AWWA taste and odor survey: Journal of American Water Works Association, v. 88, no. 4, p. 168–190.

Taylor, W.D., Losee, R.F., Torobin, M., Izaguirre, G., Sass, D., Khiari, D., and Atasi, K., 2006, Early warning and management of surface water taste-and-odor events: American Water Works Association Research Foundation Reports, 268 p.

U.S. Environmental Protection Agency, 2000, Ambient water quality criteria recommendations—Information supporting the development of State and Tribal nutrient criteria, lakes and reservoirs in nutrient ecoregion IX: U.S. Environmental Protection Agency Office of Water, Report EPA 822-B-00-011, 99 p., accessed March 7, 2008, at *http://www.epa.gov/ waterscience/criteria/nutrient/ecoregions/lakes/lakes_9.pdf*.

U.S. Environmental Protection Agency, 2009, Drinking Water Contaminant Candidate List 3—Final: Federal Register, v. 74, no. 194, p. 51850–51862, accessed July 16, 2010, at *http://www.epa.gov/fedrgstr/EPA-WATER/2009/October/ Day-08/w24287.htm*.

U.S. Geological Survey, variously dated, National field manual for the collection of water-quality data: U.S. Geological Survey Techniques of Water-Resources Investigations, book 9, chap. A1–A9, accessed April 6, 2011, at *http://pubs.water.usgs.gov/twri9A*.

Vallentyne, J.R., 1957, Principles of modern limnology: American Science, v. 45, p. 218–244.

Walker, W.W., Jr., Westerberg, C.E., Shuler, D.J., and Bode, J.A., 1989, Design and evaluation of eutrophication control measures for the St. Paul water supply: Lake and Reservoir Management, v. 5, p. 71–83.

Watson, S.B., 2003, Cyanobacterial and eukaryotic algal odour compounds: Signals or by-products? A review of their biological activity: Phycologia, v. 42, no. 4, p. 332–350.

Weete, J.D., Blevins, W.T., Wilt, G.R., and Durham, D., 1977, Chemical, biological and environmental factors responsible for the earthy odor in the Auburn City water supply: Auburn University, Alabama, Agriculture Experimental Station Bulletin 490, p. 1–46.

Welch, E.B., 1992, Ecological effects of wastewater: London, Chapman & Hall, 425 p.

Wetzel, R.G., 2001, Limnology—Lake and reservoir ecosystems (3d ed.): New York, Academic Press, 1006 p.

Wnorowski, A.U., 1992, Tastes and odors in the aquatic environment—A review: Water SA, v. 18, no. 3, p. 203–214.

Yoo, R.S., Carmichael, W.W., Hoehn, R.C., and Hrudey, S.E., 1995, Cyanobacterial (blue-green algal) toxins—A resource guide: Denver, Colorado, American Water Works Association Research Foundation, 229 p.

Young, W.F., Horth, H., Crane, R., Ogden, T., and Arnott, M., 1996, Taste and odor threshold concentrations of potential potable water contaminants: Water Research, v. 30, no. 2, p. 331–340.

Zaitlin, B., and Watson, S.B., 2006, Actinomycetes in relation to taste and odour in drinking water—Myths, tenets, and truths: Water Research, v. 40, p. 1741–1753.

Zimmerman, L.R., Ziegler, A.C., and Thurman, E.M., 2002, Method of analysis and quality-assurance practices by the U.S. Geological Survey Organic Geochemistry Research Group—Determination of geosmin and methylisoborneol in water using solid-phase microextraction and gas chromatography/mass spectrometry: U.S. Geological Survey Open-File Report 02–337, 12 p.